ElectroMagnetic Compatibility

Michel Mardiguian

ElectroMagnetic Compatibility

Understanding, Design, and Testing with Practical Solutions

Second Edition

Michel Mardiguian
St Remy-les-Chevreuse, France

ISBN 978-3-032-02687-3   ISBN 978-3-032-02688-0   (eBook)
https://doi.org/10.1007/978-3-032-02688-0

This Springer imprint is published by the registered company Springer Nature Switzerland AG
The registered company address is: Gewerbestrasse 11, 6330 Cham, Switzerland

Acknowledgments

EMC is a strange discipline and EMC engineers are a strange people. They use an esoteric language, full of bizarre terms like decibels-above-a-microvolt-per-meter-per-megahertz, BroadBand interference, or quasi-peak detection. They seem to have a divinity, Murphy, who they fear and respect. To please him, they constantly offer presents like ferrite beads or silver-plated bandages. They tell ignorant designers that if they do not follow their advice, the whole hell will break loose, and indeed, it often does....

Well, I am one of them, and while there is now a plethora of literature for EMC specialists, I have decided that the time has come to demystify the subject, in a simple parlance that any non-expert but decent electronic engineer or technician can grasp. But once the idea of such a book had popped up, it would never have stayed afloat without the timely and restless support of the editorial team at Electronic Environment Magazine, in Sweden: Dan Wallander, Leijla Musik and Thomas Glans.

For this second edition, I also want to express my gratitude to the Springer editorial group, who provided constant and efficient cooperation.

I am also indebted to the person who endured for 4 months the company of a non-professional writer who engaged in such foolish venture, dreaming of his manuscript day and night, and forgetting everything else. Thus, I want to include in these acknowledgments the one who is sharing my life: I name Corinne, my wife.

Contents

Chapter 1
What Is EMI/EMC? An Introduction to Interference Control

Preamble

Electromagnetic interference (EMI) is the generic term describing a situation whereas an electrical disturbance generated by a certain electronic/electrical equipment is causing an undesirable response to another equipment. This undesirable effect may range from a mere nuisance to a catastrophic failure, with associated financial losses or eventually human casualties. The origin of the disturbance could also be natural phenomena like lightning strokes or ElectroStatic Discharges (ESD). ElectroMagnetic Compatibility (EMC) is just the opposite: EMI being the disease, EMC is the cure, which is the discipline analyzing and preventing or fixing interference problems.

EMI has existed ever since our modern societies started using electricity for transportation, for domestic or industrial power, and for transmitting intelligence—mostly telecommunications and radio. As early as 1915–1920 where wireless transmission (Morse code or voice) started to be common place in military domain, it became rapidly obvious that several ships or army bataillons operating in same area were sometimes unable to properly communicate between each other or with their base station. In those early days of RF engineering, many EMI aspects like receiver selectivity and spurious response, channel separation, high harmonic contents of transmitters and the like were unknown.

So, little by little, some guidelines started popping out for reducing EMI situations to a tolerable level. But more rapidly than some design and fixing rules were put into practice, the astronomical growth of electronics and RF applications after the 1930s caused the number of EMI cases to skyrocket, in all domains: telephony and telecommunications, air navigation, public radio and TV services, mobile radio, etc. In the major industrialized countries, RF regulation agencies and private industry decided that the days of empirical approaches and EMI « gurus » were counted, and that a methodical strategy was badly needed. EMC was born, not as a new science, but as a multifacet discipline, and a certain number of essential aspects like source and victim dichotomy, coupling mechanisms and the like were set forth,

M. Mardiguian, *ElectroMagnetic Compatibility*, https://doi.org/10.1007/978-3-032-02688-0_1

along with specific components, instrumentation and measurement techniques, as will be seen next.

A measure of this invisible threat is given by the following few examples of the many EMI incidents that occurred worldwide in the 1970–1980, period where stringent EMC regulations were not yet applied. EMI manifests in all degrees, *from simple nuisances*:

- 70,000 claims/year (USA) and about 30,000 claims/year (major European countries) for jammed public radio and TV reception
- remote-controlled garage doors or car locking devices jammed by nearby airport radars
- video monitors jammed (image flutter) by 50 Hz power cables
- surveillance cameras jammed by nearby variable speed motor drives
- cash registers and electronic weighing machines in food stores displaying wrong figures because of ElectroStatic Discharges
- Breath analyzers (alcoholic contents) giving wrong readings when police was using their radio
- pumps in gas stations displaying wrong price when drivers were using their CBs, etc.

  to catastrophe:

- medical monitoring devices (ECG, EKG, blood analyzers) delivering wrong values when hospital or emergency squad portable transmitters and paging were used
- 30 m high crane dropping its load in harbor, when dockers were using their talkie-walkies, causing unexpected release of the crane jaws
- six ferries crashing their piers because their propeller pitch remote control was interfered by local radio-transmitters (Seattle, 1988)
- 134 deaths (seamen and pilots) and \$72 million damage due to a radar beam unexpectedly firing a missile on the deck of aircraft carrier USS Forrestal (1967)

The full list of reported incidents—not to mention the unreported ones because they were not publicized, or the EMI cause not fully demonstrated, would amount to an impressive pile (Figs. 1.1 and 1.2).

1.1 The Source/Victim Concept

Given the complexity of the interactions between the many elements that are involved, a clear, simple way for addressing the « who-does-what » of an EMI situation is the "source and victim" concept (Fig. 1.3). It states that an EMI problem can be viewed as a theater act staging three players:

- *the source of EMI*, which can be a natural phenomenon as with lightning or ESD or a man-made device that generates high frequency intentionally (authorized RF

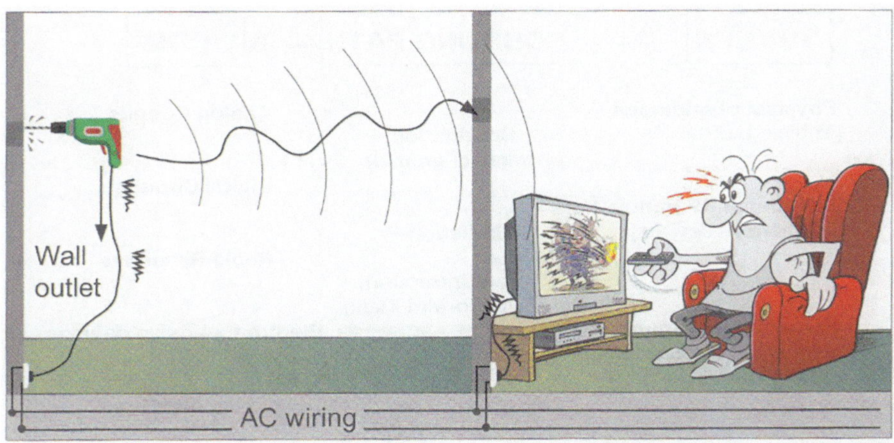

Fig. 1.1 Example of a simple, frequent EMI nuisance. The power drill is causing conducted and radiated interference to the neighbor's TV set. With analog TVs, degradation was progressive, like blurred, fuzzy picture. With digital TV, degradation is sudden, with loss of pixels, frozen image or "mosaic"

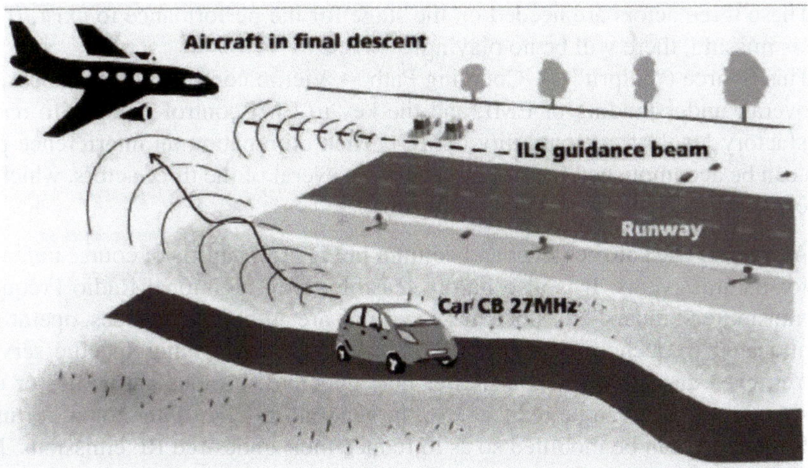

Fig. 1.2 Example of EMI situation that could end in a catastrophe. The aircraft, approaching under non-visual (IFR) conditions is guided to the runway by the ILS beam. Harmonic #4 of the CB in the vehicle on the nearby lane could interfere with the 108 MHz ILS signal, causing the aircraft to miss the runway. Although rare, such incidents do happen, and it takes a well experienced pilot to recognize the anomaly and take over the flight controls

transmitters) or as a by-product of their operation (digital circuits and switch mode power supplies).

- *the victim of EMI*, which can be any analog or digital circuit whose low level input can be activated, and eventually damaged by undesired signals.

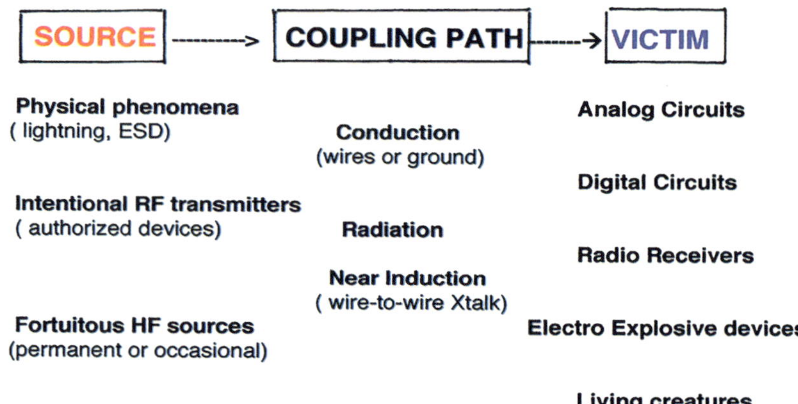

Fig. 1.3 The source-and-victim concept, a basis of the EMI/EMC strategy. In an interference case, the equipment can be the victim or the source

- *the coupling between source and victim*, which can be a conducted path, a radiated path or an in-between like cable-to-cable crosstalk.

These three actors are needed on the stage for the performance to exist. If only one is missing, there will be no playing.

This Source ("culprit") → Coupling Path → Victim combination is the basis for an overall understanding of EMI, and the key to EMI control in order to reach a satisfactory level of compatibility (EMC). Then eliminating an interference problem can be accomplished by acting on one, or several of the three actors, whichever is accessible to changes, at a reasonable cost.

1. **Acting on the source** (left-hand column in Fig. 1.3) This is of course unfeasible for natural events. It is also hardly feasible with intentional Radio Frequency transmitters, radars, etc. since these systems are authorized devices, operating at allocated frequencies and designed/installed for performing specific services. Yet, even an authorized transmitter may generate spurious harmonics or other side effects that can be reduced. So, in general only non-intentional, fortuitous RF sources can be modified so as to reduce their undesired RF emissions. Most common examples are digital circuits, switch mode power converters, pulse-driven motors, etc.
2. **Making the victim circuit less vulnerable**. This carries the constraint that the essential characteristics of this circuit like detection level, bandwidth or time constant, etc. must not be, or only slightly, affected by the change, since they are necessary to the functional performance. Therefore this option has often a very limited range.
3. **Acting on the source-to-victim coupling path**. This is probably where the designer has the largest choice of solutions: shielding, filtering, grounding, physical separation or orientation, loop reduction.

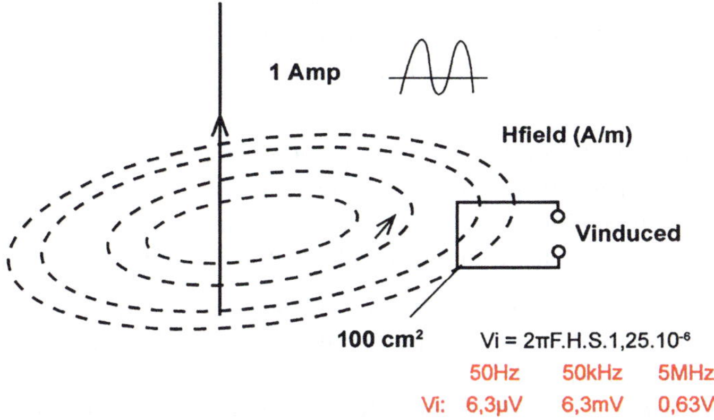

Fig. 1.4 Example of voltages induced in a 100 cm² exposed loop

Another interesting facet of this concept is that *it is perfectly reversible*. The mechanisms that could cause the circuit to be a victim are the same that could make it a potential source of interference.

1.1.1   The Importance of Frequency Aspects

Practically all coupling mechanisms that are conveying the source emissions up to the victim's input are frequency dependent: as frequency increases, the coupling coefficient increases. In some cases, it may even aggravate as frequency squared. A quick example can give a measure of this frequency aspect. Assume a *long piece of wire* carrying a constant value AC current of 1 A. At 10 cm distance, this 1 A current can cause, by magnetic field coupling, the open loop voltage shown in Fig. 1.4.

1.1.2   CE, RE, CS, RS: Four Facets of the EMC Understanding

Still based on our source–victim principle, the reduction of EMI situations to a tolerable level will be secured by controlling the EMC performance of the equipment before they are put on the market. Proper EMC tests will verify:

(a) **the Emission levels** of the equipment, when they are potential sources of interference to nearby systems, or more generally to the outside world. This is accomplished by measuring the amplitude of high-frequency noise that is caused by conduction and radiation with specific instrumentation, and then compare it to the applicable norms. These norms have been devised to protect the users of civilian or military radio spectrum.

- *Conducted Emissions* (abbrev. CE) are measured on external power cables and eventually on telecommunication cables,
- *Radiated Emissions* (abbrev. RE) are measured with calibrated antennas and special receivers,

(b) **the Susceptibility levels** of the equipment, when they are confronted to the many electromagnetic threats of their surrounding environment. This is accomplished by simulating an exposure to severe conducted or radiated disturbances, while checking that the equipment under test is still functioning properly.

- *Conducted Susceptibility* (abbrev. CS) is tested by injecting calibrated HF (sinewave) and transient pulses on external power cables, eventually telecommunication cables.
- *Radiated Susceptibility* (abbrev. RS) is tested by illuminating the equipment under test with strong fields, simulated in an EMC test Faraday cage.
- *Some specific immunity tests* like Lightning and ESD are also practiced.

The military norms have widely used this convenient (C, R, E, S) four-letter code for identifying which EMC tests are required. For instance, RE102 test is a Radiated Emission measurement covering the spectrum from AM radio band up to the upper UHF and MicroWaves.

Susceptibility or Immunity?

Generally, EMC requirements call for a *no-error or no-damage* response of the equipment under test for a given amplitude A1. On the other hand, in order to know the true *susceptibility*, the test level should be cranked-up to threshold (A2) where the equipment start showing malfunctions. The difference (A2 − A1) in dB is called the EMC margin, as being the ratio between a level that the equipment *cannot tolerate* and the *demonstrated level of immunity*. A safe practice is to have an EMC margin of at least 6, and preferably 10 dB.

1.1.3 A Summary of the Principal EMI Coupling Paths

Once understood the Culprit/Coupling Path/Victim concept, the key to understanding and preventing or solving EMI problems is to find and reduce the coupling paths that the interference will use. There are not that many, in principle:

- **CONDUCTION COUPLING**: culprit and victim are sharing a same conductor, often a ground return, for functionality safety or simply structural reasons.
- **RADIATION COUPLING**: culprit and victim are not in physical contact, separated by some interval (usually air). Interference will occur by radiation.

- **CABLE-TO-CABLE CLOSE COUPLING or Crosstalk**: culprit conductors are running close and parallel to victim wires. Interference will occur by mutual capacitive or magnetic coupling.
- **COUPLING from and to the POWER MAINS** conductors feeding the power into our equipment can carry Power line disturbances and RF interference gathered from outside, but they can also pollute the power mains by RF noise coming from our own circuits.

1.1.4   Intra-system Versus Inter-system EMI/EMC

There is an important difference between an interference that is caused to, or suffered from the environment, and the interference inside an equipment that is disturbing itself (self-jamming). These two situations are referred to as *Inter-system* and *Intra-system EMC*.

Intra-System EMC is a basic condition for making sure that the equipment is not a victim of itself. So to speak, we must insure that there is no internal source-to-victim combination that can create a performance degradation or upset. It is a designer's responsibility to check first that this is correctly addressed; otherwise the equipment will never work properly, even in a non-critical environment. On the other hand, the designer is facing a situation where he has some control on the source, the victim, the coupling path altogether.

By contrast, *Inter-system* EMC is a case where one of the "players," source or victim, is out of the designer's hands, who has no possibility of changing the environment of his system.

1.2   Practical EMC Units

Traditionally, voltage, current, and fields are expressed in Volts, Amperes, Volt/meter (E-field) or Amp/meter (H-field). In EMC when dealing with sensitive receivers or with emissions testing, these standard units are too large. So submultiples are used instead, the most common ones being:

MicroVolt (μV), MicroAmp (μA), MicroVolt/m (μV/m), MicroAmp/m (μA/m)

Example
A good FM receiver tuned on a given station has a typical sensitivity of 0.5–1 μV on its RF input (Antenna "in"). Given approximately 0.1 V per V/m for its rod antenna field conversion, what is the minimum discernable field by this radio set?

Answer: 1 μV/0.1 = 10 μV/m.

1.2.1   Why Decibels?

The Decibel is widely used in EMC community for many reasons:

- Specifications levels are most often imposed in dB
- EMC hardware (filters, shields, etc.) performances are characterized in dB
- Most measuring instruments are scaled in dB

But why is this so? Simply because an EMI situation is often facing a huge dynamic range: for instance, a victim circuit having a sensitivity in the μV or mV range may be confronted with strong fields or a power transients having amplitudes of kV. This gives 6–9 orders of magnitude to the problem. A logarithmic scaling is much more convenient than a linear one in such cases.

Also, the beauty of dBs is that, thanks to the properties of logarithms, all multiplications become additions and all divisions become subtractions.

By definition, the ratio of two powers is expressed by:

$$X\,dB = 10\log\left(P_1\,/\,P_2\right)$$

where

P_1: power in Watts (or mW) of measured or computed phenomena
P_2: Reference power in Watts (or mW)

Power is used in RF applications where power amplifiers or transmitters are the most common devices. Power is not commonly used in EMC parlance, where amplitudes are more the rule.

The ratio of two amplitudes (Voltages, currents, E-field or H-field) is expressed by:

$$X\,dB = 20\log_{10}\left(A_1\,/\,A_2\right)$$

where

A_1: amplitude of measured or computed phenomena
A_2: reference amplitude

There is more to this. The Decibel is not just used as a dimensionless figure expressing a gain or an attenuation: in EMC, we associate the dB to a unit, in order to express an amplitude, such as voltages in μV can be also expressed in dB above 1 μV which reads dBμV, currents in dBμA, and so forth (Table 1.1).

Examples

1 μV = 0 dBμV
100 μV = 40 dBμV
200 μA = 2 × 100 μA = (6 dB + 40 dB) above 1 μA, that writes 46 dBμA

A 60 dBμV RF noise, once passed through a 26 dB filter appear as:

Table 1.1 Essential amplitude and power ratios, and their dB equivalents

Amplitude ratio	Power ratio	Corresponding dB
2	4	+6
3.15	10	+10
10	100	+20
100	10^4	+40
1000	10^6	+60
0.5	0.25	−6
0.31	0.1	−10
0.1	0.01	−20
10^{-2}	10^{-4}	−40

$$60\,\text{dB}\mu\text{V} - 26\,\text{dB} = 34\,\text{dB}\mu\text{V}$$

Notice that we have subtracted dB (dimensionless ratio) to dBµV, therefore the result is in dBµV. Speaking in linear terms, we'd have divided a voltage by a number (the filter attenuation), so the result is a voltage.

When dealing with power, the Watt is often a too large unit, and the practice in radio, telecom and EMC has been to use the milliWatt, that expresses in dBm:

1 mW = 0 dBm
10 mW = 10 dBm
1000 mW (or 1 W) = 30 dBm

Converting dBm into dBµV is possible if we define the impedance where this power is applied. For instance, into 50 Ω (the most common impedance in the EMC instrumentation):

$$0\,\text{dBm}\left(\text{or}\,1\,\text{mW}\right)\text{into}\,50\,\Omega \;\; \text{corresponds to}\,107\,\text{dB}\mu\text{V}\left(\text{or}\,223\,\text{mV}\right)$$

1.2.2   Why Frequency Domain?

Except for transient pulses like power line transients, lightning, ESD, etc., EMC phenomena are most often treated in the frequency domain, because:

− Most EMC specification levels are shown on frequency scales or curves
− EMC hardware (filters, shields, etc.) performances are characterized in frequency domain
− Most measuring instruments and sensors are scaled in frequency

But why is this so?

− Many calculations (field reflection, skin effect, transfer functions, resonance, Crosstalk, etc.) are simpler to perform in frequency domain.

- Many EMI emission problems or measurements end-up in measuring at some discrete frequencies.
- For estimating the coupling of a single pulse, simple, quick calculations can be carried using a sinewave at equivalent frequency (i.e., bandwidth) reciprocal to the pulse risetime.

Therefore, in many cases where signals are known by their time waveform, the EMC specialist will translate them in frequency domain using Fourier conversion. Figure 1.5 shows the examples of frequency representations for some typical signals or pulses:

- **Caption (A)**: a pure, unmodulated sinewave. Its frequency representation is a single term at Frequency $F = 1/T$.
- **Caption (B)**: a periodic, but non-sinusoidal train of pulses. Frequency representation is a series of sinewave harmonics at multiples of the fundamental frequ. $F0 = 1/T$. In the specific case shown where pulse duration $D = 0.5\ T$, even harmonics do not exist and there are only odd harmonics: 1, 3, 5, 7, 9, etc. with amplitudes of decreasing like $1/F$.
- **Caption (C)**: same as (B), but the rise and fall times have been taken into account. Because of this, above a second corner frequency $F_2 = 0.32/t_r$, the envelope of the harmonics decrease more abruptly, like $(F)^2$. This second corner frequency is the key to many coupling mechanisms in EMC, since it represents the equivalent bandwidth occupied by a signal of which rise/fall times are known.
- **Caption (D)**: a case where the duration (D) of the individual pulses is $\ll T$. Since the period T is comparatively long, the fundamental $1/T$ is a low frequency and the series of harmonics resemble a comb with closely spaced teeth. In contrast, the corresponding spectrum extends to high frequencies because the individual pulse duration is small.
- **Caption (E)**: the specific case of an isolated pulse, like a lightning, ESD or power line transient. Since there is no repetition period, there are no fundamental frequencies or series of discrete harmonics. Instead, the Fourier series is replaced by the $\int F$ integral, and harmonics amplitude is replaced by Spectral Density, in Amp/Hz or Volts/Hz, etc., which is the reciprocal of the waveform area in (Amp $\times$ sec), or (Volt $\times$ sec), etc. The concept of the equivalent bandwidth (0.32/risetime) occupied by a single pulse help for selection of filters or shields.

Example

What is the occupied bandwidth by a digital pulse with 5 ns risetime?

Answer: $0.32/(5 \cdot 10^{-9}) = 64$ MHz

The forthcoming chapters will focus on the essential EMI coupling mechanisms, in order to:

- Understand them qualitatively, with sometimes a calculated estimate helped by simple numerical examples of real cases.
- Reduce/Control them by proper design guidelines (or fix them if too late)

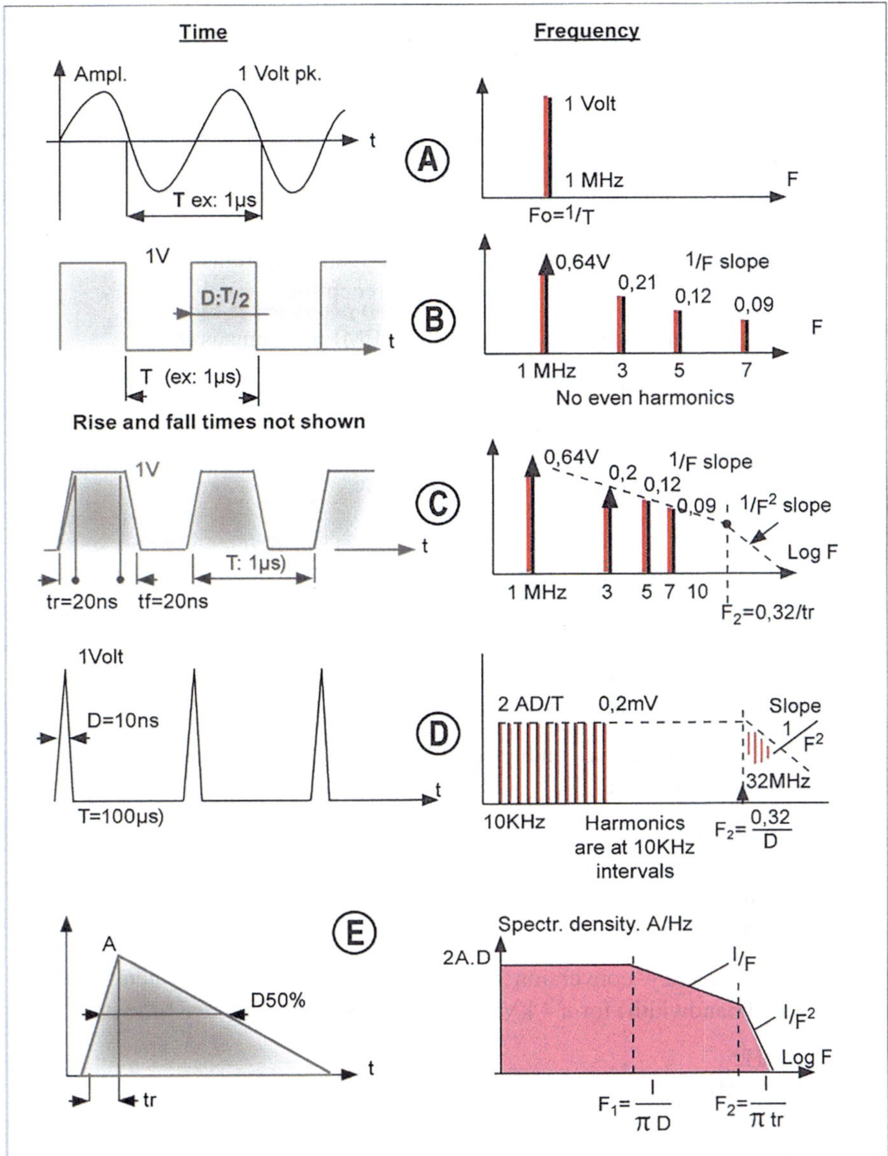

Fig. 1.5 Examples of some simple time-to-frequency conversions

We will start by paying attention to susceptibility/immunity cases, since it is often the engineer's main concern, even if this is questionable. Then we will naturally address emission problems, which will be easier because ALL the EMI mechanisms are reversible.

Quiz

There is ONE good answer or ONE that is better than the others.

1. The need for EMI control Norms and Regulations was initially urged by:

    (a) Military applications for preventing false triggering of ammunition and explosives
    (b) Civilian Organizations in the late 1930s, for reducing RF pollution of Public Broadcasting
    (c) the Safety of Air Transportation and Navigation systems
    (d) growing complaints about jammed TV reception

2. The most efficient and practical for solving EMI problems is by:

    (a) acting on the source of interference
    (b) acting on the victim equipment
    (c) acting on the coupling paths
    (d) changing the frequency of the EMI source

3. Intra-System EMC:

    (a) is the most difficult challenge because it relates to equipment reliability
    (b) is more easy to handle because the designer has more freedom for changes
    (c) cannot be treated first, because one has to comply with Inter-System EMC before Intra-System
    (d) both Intra- and Inter-System EMC are equally difficult

4. Decibels: For reducing a conducted 100 mV RF noise problem by 200 times, it has to be attenuated via:

    (a) a 46 dB filter
    (b) a 120 dB filter
    (c) a 63 dBm filter
    (d) a 23 dB filter

5. Time-to-Frequency conversion for a single pulse: the equivalent frequency (occupied bandwidth) for a 4 kV lightning pulse with 1 µs rise time is:

    (a) 1 MHz
    (b) 320 MHz
    (c) 320 kHz
    (d) 1 GHz

Chapter 2
Conduction Coupling, the First of the Coupling Paths

Preamble

As it was shown in Chap. 1, interference coupling paths can be grossly divided in CONDUCTION and RADIATION. This discrimination is purely arbitrary since no current can flow without creating an associated electro-magnetic field, and vice versa, any electromagnetic field is causing voltage and/or current to appear in exposed conductors.

However, up to a few Megahertz conduction coupling dominates EMI problems, because at such frequencies, typical internal circuits sizes and external cables lengths are smaller than wavelength λ. Wavelength and frequency are related by: λ (m) = 300/F (MHz). A 1.50 m cable length represents only $\lambda/200$ at 1 MHz and $\lambda/20$ at 10 MHz, making this cable a very inefficient antenna for radiating undesired signals. Conversely, as frequency increases above tens of MHz, cables and circuits dimensions tend to represent a progressively larger fraction of wavelength. Our 1.50 m cable now represents $\lambda/6$ at 30 MHz, where it starts acting as an efficient—although fortuitous antenna, causing conducted noise to significantly radiate.

What is addressed here is the first and major conduction mechanism: common impedance coupling (CIC). It is altogether the most easiest to understand and calculate when needed, but probably ranking highest on the scale of EMI problems in a system life. It is not necessarily the easiest one to solve because it may involve grounding at all levels: functional, safety, or structural.

2.1 Common Impedance Coupling (CIC) Mechanism

Two important facts need to be remembered here:

- Most common schematics simply address the functional aspect, representing only DC or low-frequency circuits, while real-life circuits contain high-frequency elements that are seldom shown.
- The term "ground," widely used for many applications, is often trusted as a stable reference, while it also act as a current return path, whose characteristics shall be documented, since it will carry HF currents that will not follow the classical DC electrical schemes.

A basic scheme for CIC is shown in Fig. 2.1 as a frequent cause of interference, where a great variety of circuits/equipment/systems are sharing all or part of a common conductor. Here a pulse-driven motor and an analog sensor circuit are sharing a same return wire. Problem arise because of the R × I or more generally the Z × I product in the shared segment A-B.

The conductors that are typically shared in common by several circuits are power distribution wires/busses and return or reference conductor. The latter is, by far, the most frequent with multiple-user cases, as for:

- functional return wire, for inst. The (−) wire or **Neutral** wire in power distribution
- safety conductor (Green/Yellow wire)
- structural part used intentionally as power return, like in cars, some aircrafts, and many more
- the earthing network in a building: real earth, buried earthing conductors, and many more

Looking at Fig. 2.1, let us assume the power and signal circuits have the following characteristics:

- Motor current pulses amplitude: 30 A peak, average duration 1 ms, rise time: 1 µs.
- Detection threshold of the analog monitoring circuit: 5 mV.

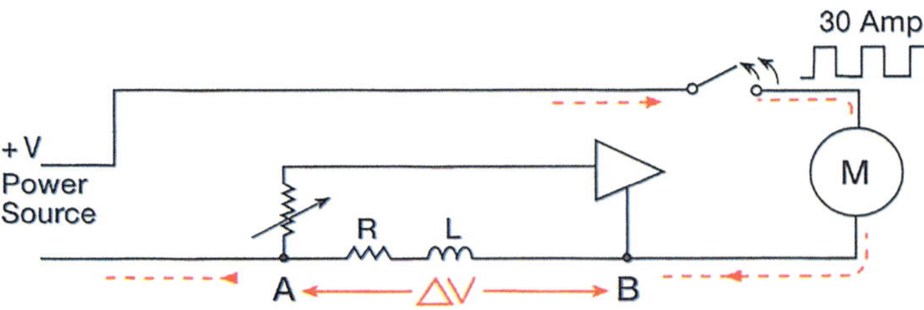

Fig. 2.1 A basic, very common case of common impedance coupling (CIC) also shown is a picture of the actual ΔV(A-B): dc view only and dc + LdI/dt spike

The common return wire is a 2.5 mm dia. with 3.5 mΩ/m dc resistance. The simple I × R product of the pulsed current along the 3 m return wire is:

$$3.5m\Omega \times 3\text{m} \times 30\text{A} \approx 300\text{mV}$$

This 300 mV ground wire shift between points A and B appears as an error voltage in series in the analog sensor loop, whose sensitivity level is only 5 mV. This is a very bad EMI situation where the noise level overrides by 60 times the detection level of the victim.

But there is more than this: the 0.3 V is just a resistive dc voltage drop. Taking into account the self-inductance of the return wire, we must add the inductive kick during the 1 µs rise time of each current pulse:

$$\Delta V = \text{L}\,\Delta I\,/\,\Delta t, \text{where L is the self\ \ inductance}$$

Ordinary round wire with diameter in the mm range has self-inductance of about 1.2 µH per meter, that is, 3.6 µH for our 3 m segment. Thus:

$$\Delta V = 3.6 \cdot 10^{-6}\text{H} \times 30\text{A}\,/\,1 \cdot 10^{-6} = 100\text{V}$$

So, not only we have a recurring 0.3 V offset that is corrupting the sensor analog signal, but on top of it, a 100 V spike appears at every current front, that can damage the analog amplifier input.

At this point, we need to introduce the term "Common Mode" voltage or current, a key definition which is the crux of many EMI manifestations and solutions. The simple circuit in Fig. 2.2 shows a wire pair carrying two sorts of currents:

(a) *the intentional current* flowing toward the load then back to its source is called Differential Mode (or "balanced") current. Looking across the wire pair, the amplitude difference between the upper and lower wire opposite currents is null, since it is the same current.
(b) *Currents coming from an outside source,* flowing on the two wires of the pair in the same direction, returning by the ground conductor (or ground, earth plane,

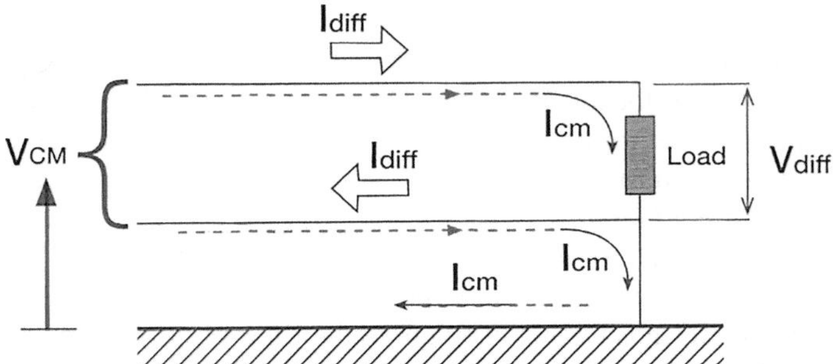

Fig. 2.2 Conceptual view of diff. and comm. mode currents

etc.) are called the Common Mode (or "unbalanced") currents. A corresponding Common Mode voltage V_{cm} is shown, as being the driving source.

Needless to say, Common Mode voltages or currents are a major cause of EMI problems, since they often originate from invisible, non-intentional sources following invisible non-intentional paths.

2.2 Basic Rules for Reducing Common Impedance Coupling (CIC)

Back to our Fig. 2.1 problem, we can now say that the A-B voltage difference along the common return conductor is a Common Mode (CM) voltage pushing a small % of the motor current in the analog sensor loop. For reducing this CIC coupling, *we must act on the $Z \times I$ product, that is, reducing Z or reducing I.* This is the basis of all the solutions we will shortly describe, summarized in a few simple rules:

Rule 1: do not allow large currents (more exactly large $\Delta I/\Delta t$) to flow in the return or reference conductors of sensitive circuits.

Rule 2: when several circuits are sharing a same reference or power distribution conductor, try making it as equipotential as possible, by lowering its impedance Z.

Applying Rule #1 means segregating the current paths, by allocating dedicated conductors to different families of circuits, such as heavy currents will not flow in the same wires or traces as a sensitive circuits' return. This is the basis of the star distribution or single point grounding (SPG) schemes. Notice that the principle applies to the (+) dc, (or phase) conductors as well, if several consumers are sharing the same power source.

2.3 Benefits and Limitations of Star (or Single Common Point) Distribution

A common way of applying Rule #1 is the star distribution principle shown in Fig. 2.3a. Functional circuits inside boxes A and B are fed by dedicated power leads, and referenced to their local 0 V ground, but isolated from their respective chassis. Thus, supply current returning from (A) to the power source on the left cannot contaminate the power return wire of box (B), and so on. It is commonly said that we have avoided a *Ground Loop*.

Figure 2.3b is another variation of Rule#1, where motor and sensor wiring has been re-arranged such as, starting from the power source, the current-hungry load is supplied first, which leads to an addendum to the Rule #1, as Rule #3 below:

Rule #3: when daisy-chaining a power distribution, always ensure that (a) the device with highest current demand is closest to the power source, and (b)

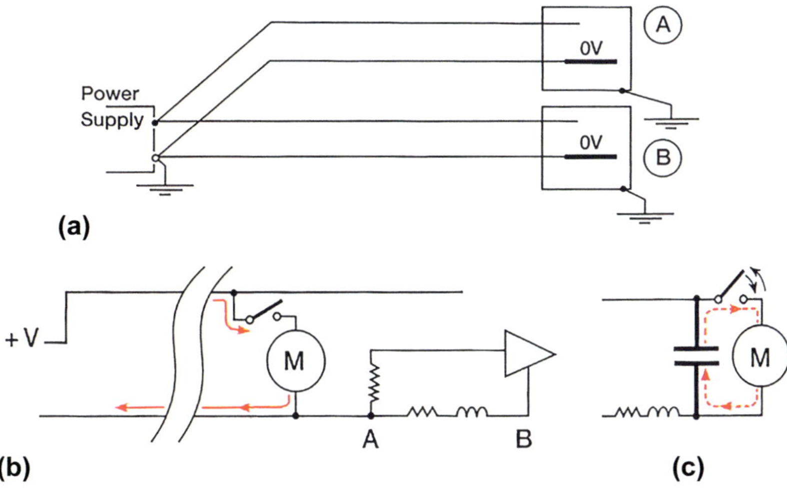

Fig. 2.3 (**a**) Practical application of Rule #1 (current segregation) by star distribution. Notice that chassis of (A) and (B) are connected to the general earthing or structural ground. (**b**, **c**) Other practical applications of Rule #1 (current segregation)

the common segment shared by multiple functions is as short and low impedance as possible.

Doing so, the large $\Delta I/dt$, going to and returning from the power source does not flow through the analog circuit wiring, hence causing no ΔV in the sensor-to-amplifier wiring. Finally, another option is shown in Fig. 2.3c, where a buffer capacitor has been added near the motor. The fast $\Delta I/\Delta t$ demands are supplied by the capacitor, acting as a local reservoir, constantly self-recharging from the battery, without creating pulsed noise on the common conductor A-B. Notice that in all these variations, we have not reduced the impedance of the return conductor. We have simply prevented large currents to flow in sensitive circuit reference.

However, there is a strong constraint for the ground loop avoidance to work that translates in two additional rules:

Rule 4: If Single Point Ground (SPG) is chosen, circuits (A) and (B), (C), and so on must have only ONE common node for their (+) and (−), at the power source.

To achieve this, the dedicated 0 V references of A, B etc... must be isolated from their local chassis if, as often the case, this chassis is locally grounded for safety or mechanical reasons. Violating this rule may cause circulation of some % of return current (A) in the 0 V reference of (B), as shown in Fig. 2.4a. The resulting ground loop, easily checked with a current probe, is that some power supply return current of A is now crossing the 0 V reference of B to return to M, the Power Supply SPG. This unwelcome current flow may disturb the operation of (B) sensitive circuits by Comm. Impedance Coupling.

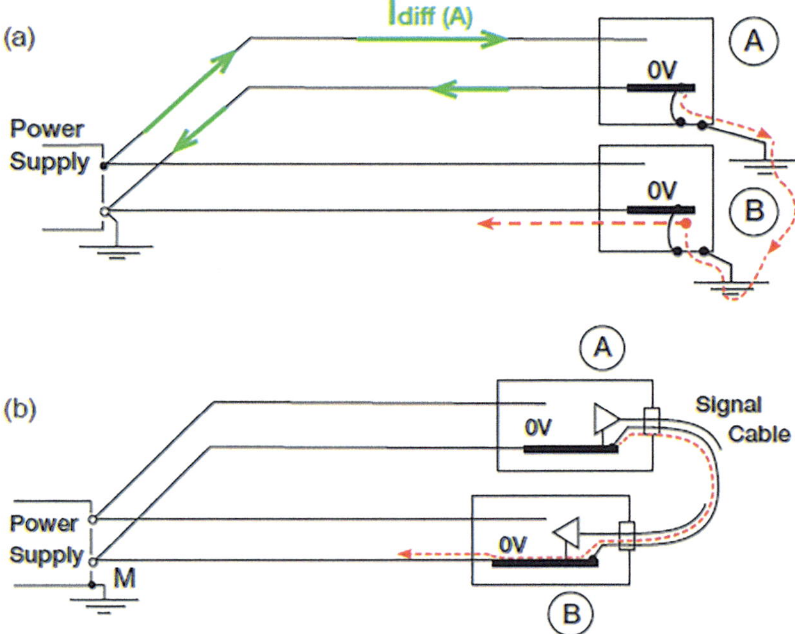

Fig. 2.4 Violation of the single point ground principle by (**a**) not isolating the signal ground (0 V) from chassis or (**b**) interconnecting the 0 V of boxes A and B via a signal cable. Some of the box A current, flowing through box B signal reference, is closing back to the Power Supply ground M

Rule 5: Boxes at the ends of a star network must not exchange directly signals, except through isolated links.

On Fig. 2.4 caption (b), we see another case of compromising the SGP purpose and creating a ground loop. Boxes A and B are exchanging data by an ordinary signal cable. Each end of the link is referenced to its local 0 V. Therefore, a new sneaky, parallel path is created for the return current of A to close back to SPG "M" via the 0 V wires in the signal link and the equipment B.

Figure 2.5 shows solutions to accommodate the above violations without creating ground loops: On the power input, a transformer is interrupting the primary loop, such as the designer has now the possibility to connect its electrical 0 V to the equipment chassis. The I/O port where A and B are exchanging signals is equipped with small signal transformer or optical isolator which restore the necessary isolation, hence the SGP principle.

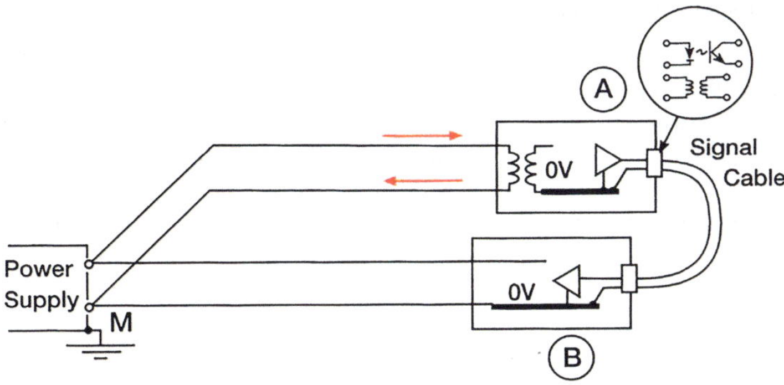

Fig. 2.5 Restoring the single point ground principle by a primary isolation device (transformer or dc-dc converter) on power input and by using signal isolation transformers or optical couplers on the I/O ports. Now, box A could have its 0v reference grounded to chassis, without creating another ground loop

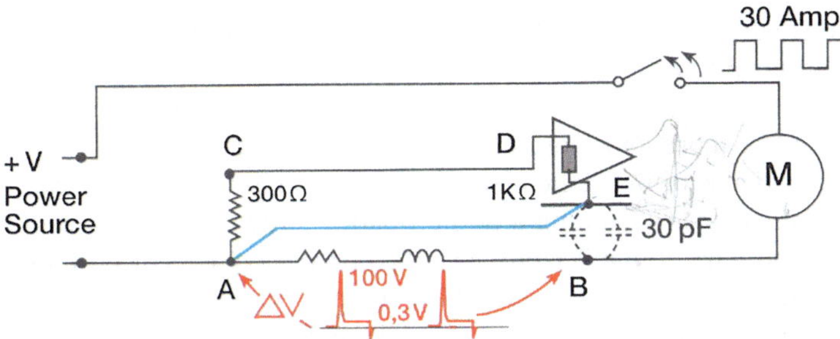

Fig. 2.6 A low frequency representation assumes that the sensor ground E-A wire is practically a short-circuit, such as the PCB 0 V (pt. E) and power source 0 V terminal (A) are ≈ equipotential. But when frequency increases, the impedance of the E-A 0 V return wire increases, while the stray capacitance of the PCB makes the floating impedance E-B progressively lower

2.4   High-Frequency Limitations of SGP (Star) Principle

In the CIC example shown in Fig. 2.1, we found a ΔV of [0.3 V + 100 V] spike appearing in series in the common return segment A-B. However, this noise voltage does not appear entirely at the amplifier input. We have to account for the divider ratio in the sensor-to-amplifier loop:

$$\text{Vinput} = V_{(A\ B)} \times 1000 / (300 + 1000) = 0.77 V_{(A\ B)}$$

Thus, because of the high input resistance of the amplifier, practically ALL the CIC voltage appears at the amplifier input. Changing to an SPG (star) scheme, Fig. 2.6 shows a dc/low frequency solution: the ground loop seems now open at amplifier low side. The floating point E sees the CIC voltage at point B through an

Table 2.1 Ground loop coupling with example shown in Fig. 2.6

Frequency	100 kHz	1 MHz	10 MHz
Impedance of 30 pF cap.	32 kΩ	3.2 kΩ	320 Ω
Impedance of 3 m wire	2 Ω	20 Ω	200 Ω
CM rejection at pt. E	$0.6 \cdot 10^{-4}$ (-84 dB)	$0.6 \cdot 10^{-2}$ (-44 dB)	0.6 (-4 dB)
Peak voltage at amplifier input	5 mV	0.46 V	46 V

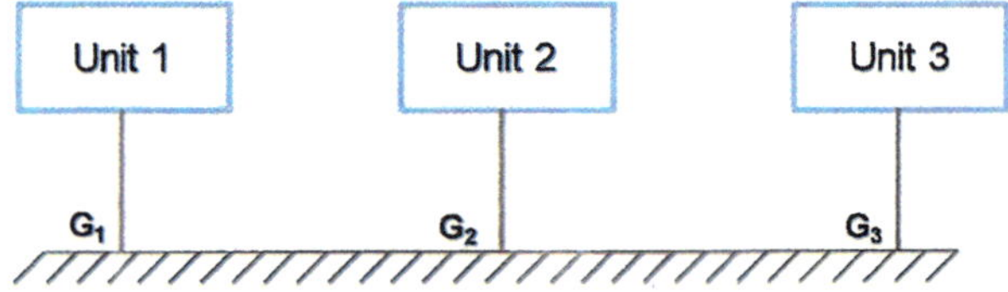

Fig. 2.7 Principle of a multipoint grounding (MPG), compared to SGP

infinite impedance, and NO voltage appears at the victim input. The CM rejection V_{A-B}/V victim is infinite.

Yet, as frequency increases, no circuit can be totally isolated: the stray capacitance at pt. E let more and more current flowing in the isolation gap. At the same time, the impedance of the dedicated sensor grounding wire E-A increases, because of its self-inductance. Assuming a 30 pF floating stray capacitance and 3.6 µH for the 3 m ground wire, Table 2.1 provides the progressive loss of ground loop isolation when frequency increases. Numbers indicate how much actual reduction is provided by the isolated scheme. At 10 MHz and beyond, it is clear that the SGP becomes totally inefficient, not to mention a strong parasitic resonance at 15 MHz. We will see further how to cope with these limitations.

2.5 Reducing CIC Via Multipoint Ground

We said before that reducing the (Z × I) product can also be achieved by reducing the value of shared impedance Z, and tying directly all sub-assemblies/equipment of a system to a same return conductor. This application of Rule #2 means that we accept to mix all return currents in a same conductor (Fig. 2.7).

The price to pay is that this conductor must have low enough dc resistance and HF impedance for the resulting (Z × I) voltage to be tolerable for the most sensitive users of this shared ground. Unfortunately, above a few kilohertz, any linear conductor: round wire, flat strap or braid, PCB trace, exhibit a self-inductance that will rapidly make it unable to act as equipotential ground.

Therefore we must change 1D shape to 2D shapes, like a plane or a grid. A large plane has practically no self-inductance (see Fig. 2.8) if it is sufficiently large (no trace or wire near the plane's edges), and can be used for multipoint grounding of various elements. Notice that resistance or impedance of a plane (sheet metal, PCB

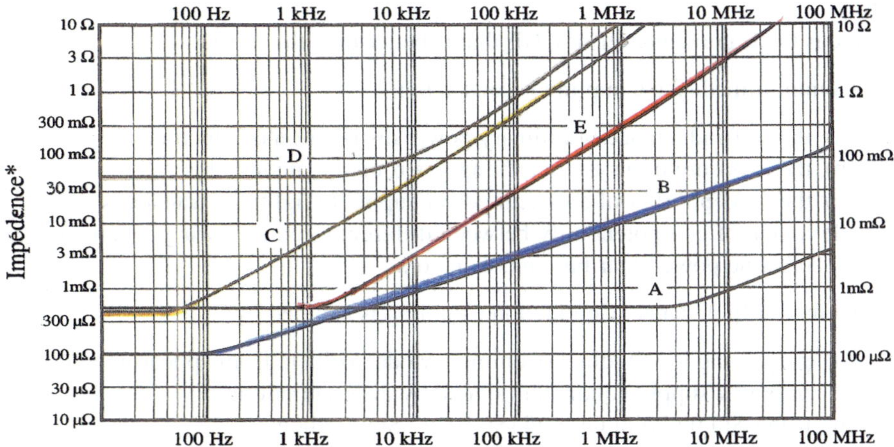

Fig. 2.8 DC resistance and HF impedance of various conductors. (**A**) Copper layer 30 μm thick, (**B**) ordinary steel sheet 1 mm thick (planes impedance given in Ω, or mΩ, per sq. area). (**C, D**) 1 m long copper wire, respectively, 6 and 0.6 mm. (**E**) Copper strap 0.3 × 10 × 100 mm. Do not use graph values for frequencies F(MHz) >60/length

copper layer, etc.) is given in **Ω/square**. Whatever it is mm^2, sq. cm^2, inch^2, m^2, etc., the value is the same.

This is what is done in *multilayer PCBs*, with Vcc and 0 V copper planes, and in RF devices (Txmitters, Receivers, amplifiers, mixers, etc.) where the cast aluminum or zinc-plated box is normally used as the RF signal reference, including for connection of the coaxial cables.

2.5.1 *Summary on Common Impedance Coupling (CIC)*

- CIC is one major cause of conducted EMI problems. It can even contribute to Radiated EMI when a noisy ground excites I/O cables.
- The prime cause of CIC being a Z × I product, an efficient solution it is controlling **I** or **Z**.
- Reducing I in a conductor shared by several very different classes of circuits implies managing the current paths. A most common way being the SPG (or star) distribution for power and ground.
- SPG becomes inefficient when branching conductor length exceed λ/10 or 30/F(MHz).
- Multipoint (MPG) ground, achieving an equipotential reference is a better answer at any frequency, provided a good ground plane can be used (ex: PCBs).
- If CIC cannot be reduced, other techniques like differential links, EMI filters, cable shielding can be used, allowing an equipment *to function even in the presence of a CIC*. They will be addressed in forthcoming chapters, since they are a cure for many kinds of EMI coupling.

Quiz

There is ONE good answer or ONE that is better than the others.

1. Comm. Impedance Coupling is caused by:

   (a) the self-inductance of ordinary power distribution conductors
   (b) the exposure of a system cabling to strong magnetic fields
   (c) the susceptibility of a system power supply to lightning transients
   (d) a lack of equipotentiality between different circuits sharing a same ground or same power distribution conductors

2. Common Impedance, Conduction Coupling is avoided by:

   (a) preventing longitudinal (Common Mode) Voltage from appearing in ground or power supply conductors shared by different circuits
   (b) using Single Point Ground (SGP) or "Star" scheme
   (c) using Multi Point Ground (MPG) scheme
   (d) Increasing the diam. of power and/or return conductors

3. A 3 A current change with 100 ns risetime flowing along 2 m of ordinary power distribution wire will cause a transient ΔV spike:

   (a) 0.3 V
   (b) 60 mV
   (c) 300 V
   (d) 60 V

4. When several boxes of a system are interconnected in a "star" scheme, CIC is prevented by:

   (a) adding direct inter-box ground wires to equalize between the tips of the star branches
   (b) grounding locally the 0 V references of all the "slave" boxes
   (c) floating locally the 0 V references of all the "slave" boxes
   (d) floating efficiently the slave boxes via isolation transformers or optical devices, or use differential links

5. A Single Pt Ground (SGP) scheme becomes inefficient for preventing C.I.C when:

   (a) the length of the branching conductors exceed $L(m) = 30/F(MHz)$
   (b) direct inter-box links are added between the branches of the star
   (c) the isolation of the 0 V references vs. chassis ground is bypassed by the stray capacitances
   (d) all or any of the above

Chapter 3
Radiation Coupling

Preamble

The former chapter treated the frequent mechanism of Common Impedance Coupling, one that probably comes at the top of the list for causing EMI problems. This chapter is addressing coupling paths where the culprit and victim circuits are not in physical contact, the parasitic effect occurring through radio propagation.

3.1 Field-to-Cable Coupling

Although neither regarded as radio antennas nor designed for, cables and printed circuit traces **are** unintentional antennas. Any piece of conductor carrying a current (or excited by a voltage) will radiate an electromagnetic field. Reciprocally, any piece of conductor illuminated by an ambient electromagnetic field will exhibit a current flow and a related voltage. Such fortuitous antennas can be broken-down in two simple forms: closed loops or open-ended wires (Fig. 3.1).

A loop is intuitively regarded as magnetic field (H-field) sensitive, but in fact it responds to both E- and H-fields, with a predominant ability to privilege one or the other depending on its orientation versus the field vectors and propagation. Open-ended wires are generally regarded as E-field sensitive, yet both shapes can pick-up signals from an electromagnetic field, defined in V/m.

Most often, real-life circuits and cabling configurations are neither purely open wires or perfect loops, but somewhere in-between. Circuits terminated into high impedances ($>377\ \Omega$) tend to behave as dipoles or whip antennas, while those terminated in low impedances ($<377\ \Omega$) tend to behave as loop antennas.

Three factors are playing a role in the efficiency of fortuitous antennas for picking-up an ambient, time-varying field (alternating or simply pulsed):

– the dimensions of the loop or wire

© The Author(s), under exclusive license to Springer Nature Switzerland AG 2025

M. Mardiguian, *ElectroMagnetic Compatibility*,
https://doi.org/10.1007/978-3-032-02688-0_3

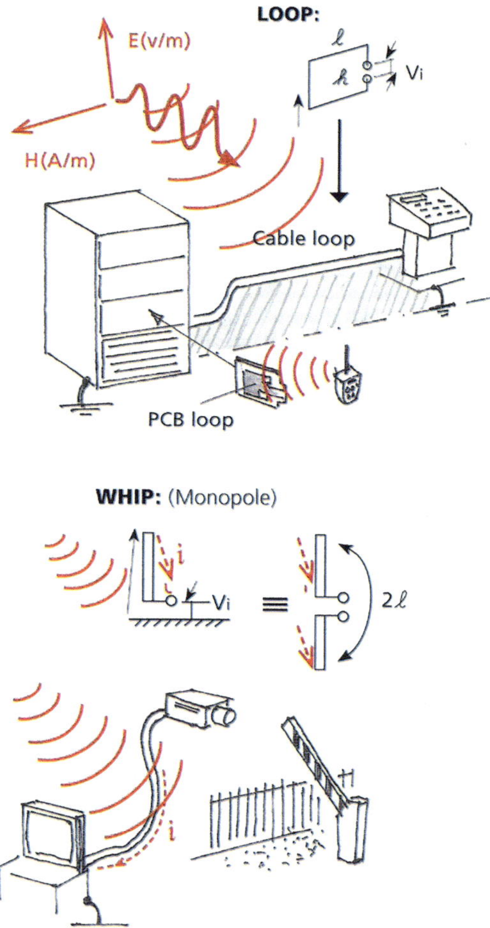

Fig. 3.1 Basic forms of field-sensitive or field-radiating shapes

- the frequency (or rise time for a pulse) of the radiated threat
- the amplitude of the field.

One good measure of the ability of the unintentional antenna to convert V/m into V is by comparing its physical length with the incident field wavelength « λ ». Given that the relationship of λ to frequency is:

$$\lambda(m) = 300 / F_{(MHz)}$$

we know that a wire (or loop) whose dimension is reaching λ/2 (λ/4 for a whip) has its maximum aptitude for field capture. But even below this ideal resonance, any conductor of length «l» has a pick-up efficiency proportional to its 2*l*/λ ratio. At the frequency of 10 MHz, that is for λ = 30 m, a wire whose length is only 1 m starts being a fairly able antenna since it represents 1/15th of a perfectly tuned dipole.

Few examples of frequently found incidental antennas are shown on Fig. 3.2. Notice that in (a), the loop is NOT the intentional path for the link. The cable can be a bundle of many signal or power wires, including their respective return (or «ground») conductors. The voltage induced in the large loop by the incident field is a Common Mode (CM) voltage appearing in the loop that tends to **push a current in the same direction in all the wires** of the pair or bundle. We have already seen such kind of parasitic voltage developing in Chap. 2 "Comm. Impedance Coupling" (CIC).

How much of it actually reaches a sensitive circuit at either end of the link depends on the way the receiving circuits are balanced (symmetrical) or not, or if the cable is shielded or not. A line-to-ground CM voltage, if large enough like hundreds or thousands of volts can also simply destroy sensitive, unprotected circuits.

It is worth noticing that this voltage relates to the flux linkage in loop the physical area (m^2), *REGARDLESS the loop is electrically closed or not*. Simply, if the loop is open, like floating one of the 0 V reference of the electronic circuit, in theory no current could flow in the cable. But we have already seen (Chap. 2) that such *star* or *single-point* grounding solution, progressively loses its efficiency when frequency increases, because of the parasitic capacitances to ground in the equipment.

In captions (b) or (c) the differential mode (DM) loop is of smaller size, like two wires of a same pair, or PCB traces, forming a loop with a few tens cm^2 area. But this is a differential scheme, where the picked-up voltage can directly upset the analog or digital ICs attached to the pairs.

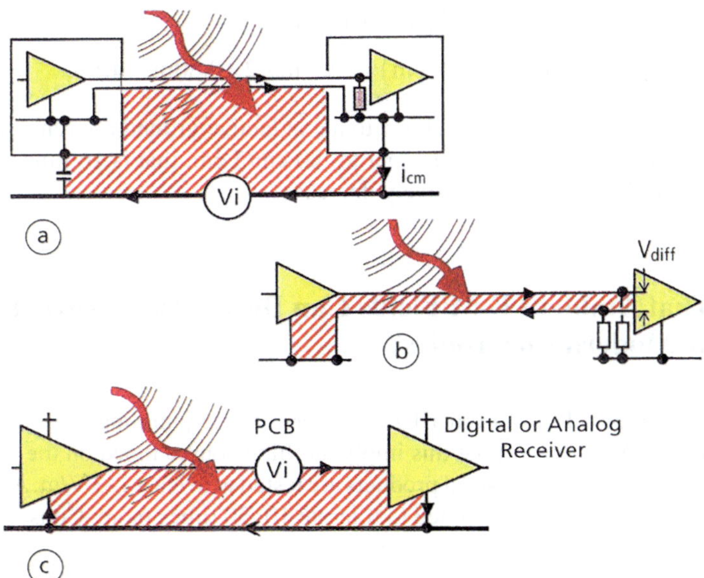

Fig. 3.2 A few examples of unintentional antennas

Three simple formulas give a good estimate of the worst-case loop-induced voltage Vi:

$$\text{for } F(\text{MHz}) < 100/1(\text{m}): Vi = 1.h(\text{m}^2) \times E(\text{V}/\text{m}) \times F(\text{MHz})/50 \quad (3.1)$$

$$\text{for } F(\text{MHz}) \geq 100/l(\text{m}) \text{ and } < 100/h(\text{m}): Vi = 2h(\text{m}) \times E(\text{V}/\text{m}),$$
$$\text{independent of } F \& l \quad (3.2)$$

$$\text{for } F(\text{MHz}) > 100/h(\text{m}): Vi \approx 120/F(\text{MHz}) \quad (3.3)$$

with l and h being the loop length and height

Figure 3.3 is showing field-to-loop coupling factor for a few simple sizes, along with the E-field pick-up ability for a typical "whip" antenna configuration, where the base of the antenna can be a grounded equipment while the I/O cable is connecting a small, totally isolated access-control device, surveillance camera or alarm device, grounded to nowhere.

Numerical Example

What is the voltage induced by a 3 V/m ambient at 100 MHz on a 1.50 m cable 0.80 m above ground?

For this geometry:

- Equation (3.1) is valid up to F = 100/1.50 = 66 MHz
- Equation (3.2) applies from F ≥ 66 MHz up to F = 100/0.80 = 125 MHz
- Equation (3.3) applies for F ≥ 125 MHz

Here, 100 MHz is >66, thus Eq. (3.2) applies:

$$Vi = 2h(\text{m}) \times E(\text{V}/\text{m}) = 3\text{V}/\text{m} \times 2 \times 0.80\text{m} = 4.8\text{V}$$

If the cable could be brought closer to a conductive ground reference, like 5 cm for instance, the benefit would be equal to: 80 cm/5 cm, that is 16 times or 24 dB, up to ≈200 MHz, then it progressively decrease.

3.2   What Ambient E-M Fields Can Be a Serious Threat for Modern Electronics?

Letting aside radio, TV, and other RF receivers, where jamming can occur at very low levels (1 mV/m or less) but this imply an illicit source tuned on the same radio station, most consumer/industrial products are tested to at least 1–3 V/m. Automobile or airborne electronics are designed to tolerate 10–30 times more. Consequences of a field exposure also depend on:

(a) *its frequency*; most critical frequency range is in the tens to thousands megahertz, since related wavelengths are 30 m down to 0.30 m, respectively. Thus cables as short as a meter are already efficient pick-up antennas.

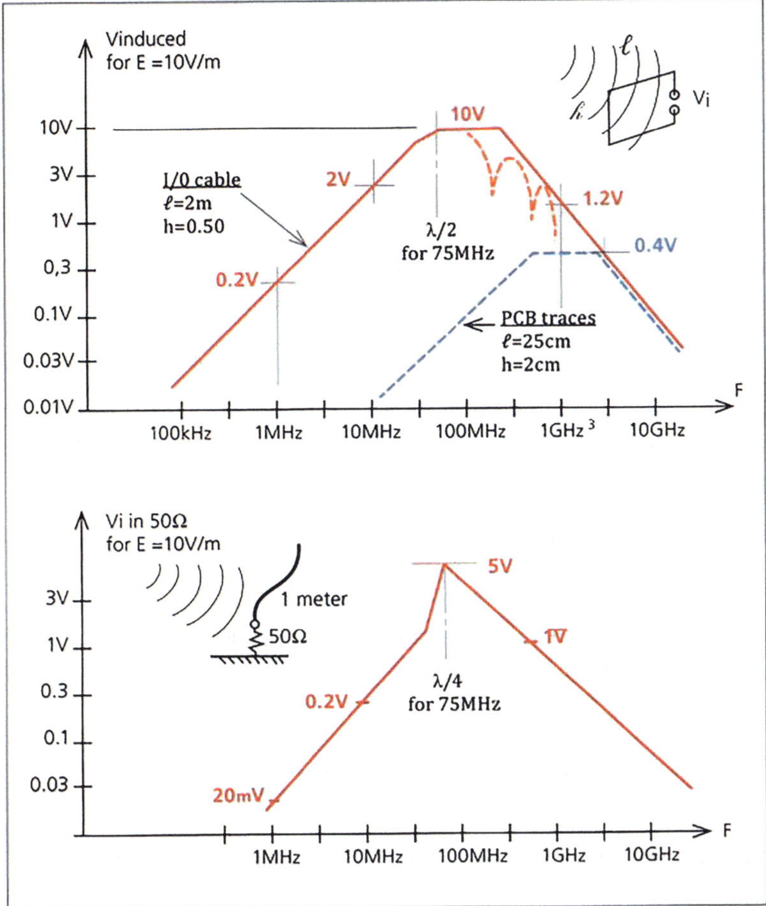

Fig. 3.3 Field-to-loop coupling factor for a few simple shapes, given as open-loop voltage for a constant 10 V/m field, a severe environment (industrial) in non-military applications. For the monopole (whip-like), voltage is given in a 50 Ω load at the base of the open-ended cable

(b) *its quasi permanent, or transient nature*: A 50 kA lightning strike at 30 m distance will cause a 100 kV/m field, that is a tremendously high value. But if the installation is reasonably protected, the consequence may simply be a temporary upset, which can self-recover or at worst require a reset or re-power-up. On the opposite if a constantly occurring RF field is causing errors, the system will be down all the time.

   Table 3.1 provides an idea of common radiated exposures, with their frequency and distance as variables.

Table 3.1 Some E-M field exposures. For hi-gain antennas, field given in boresight of main beam

Source	Frequ. (typ)	Equivalent rad. power (typ) incl. antenna gain if any	Field (max) at distance	Vinduced, for Emax
AM Broadcast.	0.15–1.6 MHz	50–1000 kW	5 V/m at 1 km	0.16 V in 1 m² loop
FM Broadcast	88–108 MHz	100 kW	1.8 V/m at 1 km	3.6 V in 1 m² loop
UHF TV	470–853 MHz	300–500 kW	3.5 V/m at 1 km	0.5–1 V in 1 m² loop
Radars	0.4–40 GHz	kW to 10 GW		
CB on vehicle	27 MHz	12 W	6 V/m at 3 m	3 V in 1 m² loop
Hand-held radio	145 MHz	2–12 W	6 V/m at 3 m	18 V in 1 m² loop
Cell phone	900 and 1800 MHz	2 W (max)	6 V/m at 1 m	0.12–0.25 V in 10 cm²
Lightning 50 kA	Rise time: 1 µs		100 kV/m at 30 m	300 V in 1 m² loop
ESD 5 kV, personal	Rise time: 1 ns		10 kV/m at 0.1 m	4 V in 1 cm² loop

How does one know the ambient field from a given RF transmitter if no data are available? A simple formula can give a coarse approximation of the E-field in V/m at a distance D:

$$E = \frac{1}{D}\sqrt{30\mathrm{ERP}}$$

(3.4)

with, ERP: effective radiated power (W) = Txmitter Power × Antenna Gain[1]

3.3 What Are the Solutions Against Field-to-Cable Coupling

There are many solutions for reducing the effects of a radiating source that is assumed beyond our control. One can reduce the dimension of the victim's capture area, shield the conductors or circuits that behave as receiving antennas, or filter the RF currents before they reach the victim circuits inside the equipment.

As we did for the CIC, next are several simple rules against the effects of radiation coupling, that are not contradictory with those for conduction coupling. In fact, both are complementary, since conduction problems manifest most at frequencies below tens of megahertz, a domain where radiation coupling is usually weak, and vice versa. This is why these rules are numbered in sequence after those of Chap. 2.

---

[1] Numerical Gain (not in dB).

Rule #6: Reduce by all means the geometrical dimensions of the capture areas existing wire-to-wire (or trace-to-trace) and cable-to-ground.

- For installations, try to run the external cables as close as possible to a metallic structure or reinforced concrete slab; the steel grids or rebars make a decent HF reference, reducing the induced CM voltage.
- For wire pairs, keep close to each other the positive and return wires of a same signal. Twist them together; but this reduce Diff. Mode pick-up only, not the CM. For PCBs, keep the positive and return traces close (side-by-side) or on top of each other. It reduces induced Diff. M voltage.

Rule #7: if you cannot bring the conductor down to a ground plane, bring a ground plane up to the conductor.

- For installations, run the cables in solid or perforated metal raceways, making sure all their elements are continuously bonded/screwed from end-to-end and to the equipment enclosure. A large "U" shaped raceway enveloping the cables acts as a substitute to ground plane.
- For installations without accessible metal ground plane, install a large (at least 2 m^2) metal plate underneath the equipment cabinets, extending beyond the equipment footprint.
- For PCBs, use multilayers with plain (no slots) ground planes.

Rule #8: Use shielded cables with metallic connectors making a tight, peripheral contact with the shielded jacket.

Install them with a good, direct contact of the shield at the penetration in the metal enclosure of the equipment. If the equipment has no metallic (or metallized plastic) enclosure, ground the I/O cables connectors to the main PCB ground plane, as close as possible to the point of entry.

Rule #8.a: Cables shields must be connected directly to the equipment, both ends.

Stay away from old, die-hard myths that say « never ground a shield both ends, because this creates ground loops ». Open-ended shields do not provide an alternate path for cable-loop current, therefore there would be no change in the CM current flow on your signal link. The only case where grounding a cable shield both ends could be detrimental is when a large, low frequency (50 or 400 Hz) ground shift exist between two equipment on a site. But in this case, the shield with one end floated remains worthless against field coupling into the cable loop *at any frequency*. Only low frequency capacitive (E-field) coupling will be attenuated.

Rule #9: If none of rules 6, 7, 8, 9 is feasible or controllable, let the antenna (capture loop) exist and block the RF currents by EMI filters at the I/O ports.

This is somewhat similar to what has been explained in Chap. 2 for C.I.C., when discussing the merits and limitations of floated 0 V references (single-point ground schemes).

The common mode-induced voltage in a cable-to-ground loop does not appear entirely at the victim's receiver input. We have to account for the divider ratio of the victim's input resistance to the total loop impedance, like in the example of Chap. 2, Table 3.1.

The ratio (Vict imput impedance)/(Total loop impedance) is called CM-to-Diff-Mode conversion. Like for CIC and beside shielded cables it can be controlled by SGP schemes (only against low frequency), victim input filtering, CM ferrite toroids and isolation transformers.

Quiz

There is ONE good answer or ONE that is better than the others.

1. The voltage induced by an ElectroMagnetic field on cables:

   (a) exist only if the cable-to-ground loop is open at one end
   (b) is maximum when cable length reaches $\lambda/2$
   (c) is maximum when the cable height above ground reaches $\lambda/2$
   (d) increases if the victim has a high impedance input

2. Twisting a wire pair for controlling CM field pick-up:

   (a) reduce the CM field pick-up at any frequency
   (b) is efficient against CM only at low frequencies
   (c) is useless
   (d) is always efficient because it forces the CM current to flow in opposite directions

3. For reducing the field pick-up by a cable:

   (a) keep the signal wires far from the common ground conductor
   (b) make sure the system boxes are connected to a good, low-resistance earth rod
   (c) bring the signal wires close to a ground plane or use a metallic raceway, or shield the cable
   (d) orient the cables such as it is parallel to the direction of magnetic field lines

4. What is the **best** method to reduce field Comm. Mode coupling into a cable OR the Comm. Mode-to-Diff. Mode Conversion?

   (a) twist the wires or add a screen on the pairs
   (b) use optical isolators
   (c) Use differential (symmetric) transmission or float the terminal circuits or use opto-isolators
   (d) None of these

5. What is the E-field received at 100 m from a 100 MHz FM transmitter with output power 10 kW and antenna gain 6 dB (that is a numerical gain = 4)?

   (a) 0.3 V/m
   (b) 11 V/m
   (c) 33 V/m
   (d) 1.8 V/m

Chapter 4
Cable-to-Cable Coupling or Crosstalk

Preamble

The term Crosstalk, inherited from the early days of telephony, describes a case where during a conversation between two subscribers, one could hear the talk of a third party. Technically speaking, Crosstalk is a near-induction mechanism by which two wires, belonging to different circuits and running in parallel can couple transversally, a certain % of the « culprit » wire voltage appearing on the « victim » wire.

Since the two involved circuits are not in physical contact, one could see it as radiation coupling. However, there is a difference: with radiation, the source circuit is emanating a field in a 3D space, where it can be received by anybody. In Crosstalk, the coupling occurs in the limited space of the wire-to-wire capacitance or mutual inductance; no RF propagation is involved.

4.1 Basic Crosstalk (Xtalk) Description

As shown in Fig. 4.1, Crosstalk (Xtalk) happens because of the distributed capacitance C_{1-2} (Capacitive Xtalk) or mutual inductance M_{1-2} (Magnetic Xtalk) that exists between two wires belonging to different circuits. This can also take place between two traces on a PCB or ribbon cable, between two pairs (Diff. Mode Xtalk) or between two cable bundles above a ground plane (Common Mode Xtalk).

Crosstalk is governed by several parameters belonging to the circuits' configuration. It aggravates when:

(a) Ratio h/s increases.
(b) Both conductors are embedded in a dielectric (more capacitive Xtalk).
(c) Victim impedance is high (more capacitive Xtalk).
(d) Culprit impedance is low (more primary current, hence more magnetic Xtalk).
(e) The parallel run of victim/culprit wires is longer.

© The Author(s), under exclusive license to Springer Nature
Switzerland AG 2025
M. Mardiguian, *ElectroMagnetic Compatibility*,
https://doi.org/10.1007/978-3-032-02688-0_4

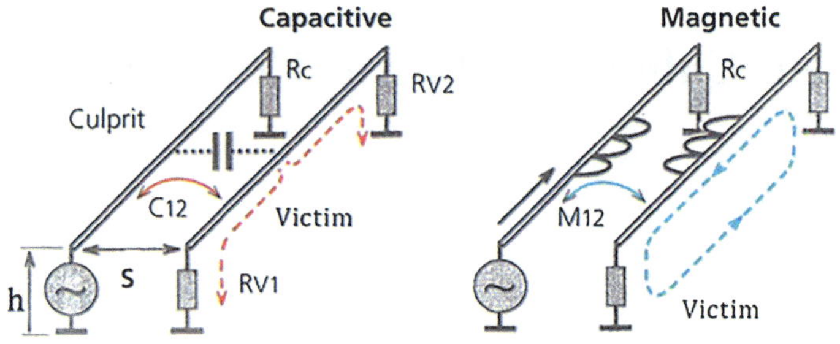

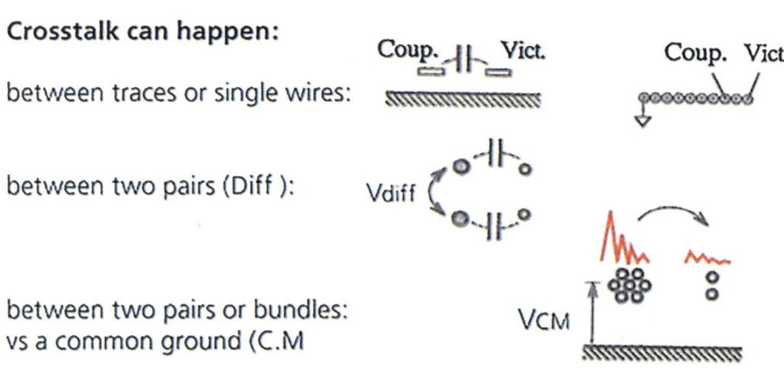

Fig. 4.1 General configurations of cable-to-cable coupling. Looking at the currents directions, notice that Cap. and Magn. Xtalk are adding on culprit source side, and subtracting on the opposite end

Crosstalk is defined as the ratio of the victim-induced voltage to the culprit voltage:

$$\text{Xtalk} = V_{\text{victim}} / V_{\text{culprit}}, \text{ that is a} \left(\text{Volt} / \text{Volt} \right), \text{ dimensionless ratio.}$$

It can be expressed in frequency domain, for sinewave signals, or in time domain for pulsed signals. The frequency diagram Fig. 4.2 shows that Xtalk never exceed or even reach a 100% ratio (that is 0 dB): victim's induced voltage cannot equal or exceed the culprit voltage.

When expressed in frequency domain (sinewave situation like a single RF interference or a discrete harmonic of a digital signal):

$$\text{Xtalk coefft} \left(\text{Capacitive} \right) = R_v / \left(R_{\text{vict}} + 1 / C_{1-2}\omega \right) \approx R_v C_{1-2}\omega$$
$$\text{if } R_v < 1 / C_{1-2}\omega \tag{4.1}$$

$$\text{Xtalk coefft} \left(\text{Magnetic} \right) = M_{1-2}\omega / R_{\text{culp}} \text{ if } M_{1-2}\omega < R_{\text{culp}} \tag{4.2}$$

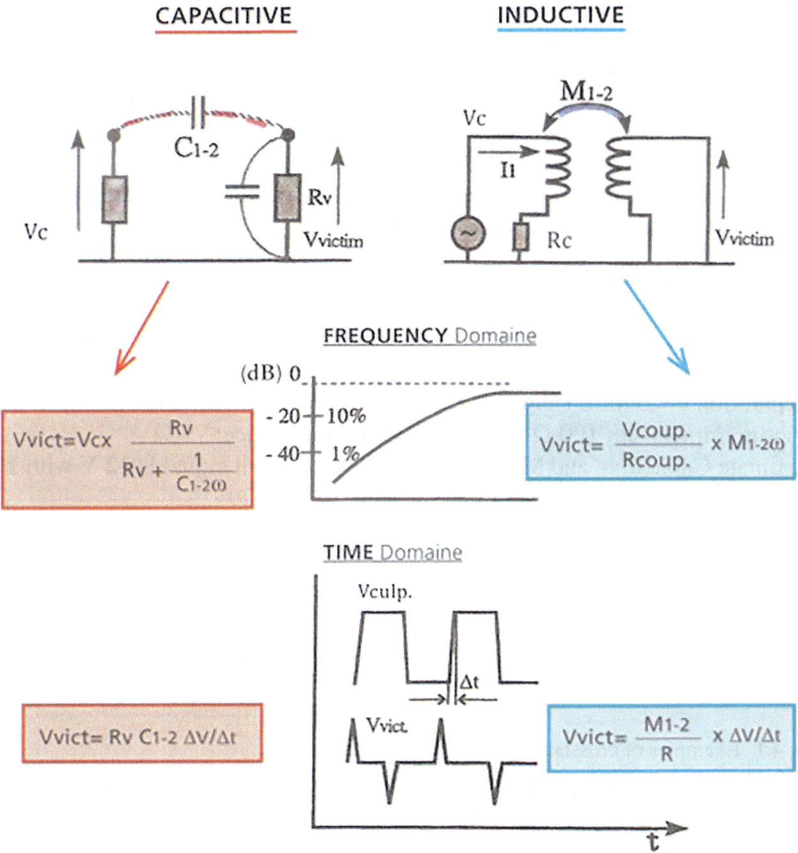

Fig. 4.2 Simple equivalent circuits (represented as end-view) for Capacitive and Magnetic Xtalk

with,

C_{1-2}, M_{1-2} = wire-to-wire coupling capacitance and mutual inductance
R_{vict} = parallel combination of victim near-end and far-end resistances (R_{v1}//R_{v2})
R_{culp} = culprit circuit resistance
$\omega = 2\pi F$

In time-domain representation (when the rise time Δt of culprit signal is known):

$$\text{Capacitive Xtalk coefft} = R_v C_{1-2} / \Delta t, \quad \text{if } R_v C_{1-2} < \Delta t \tag{4.3}$$

$$\text{Magnetic Xtalk coefft} = \left(M_{1-2} / R_{culp} \right) / \Delta t, \quad \text{if} \left(M_{1-2} / R_{culp} \right) < \Delta t \tag{4.4}$$

Notice that the capacitive Xtalk is governed by the victim's resistance, while magnetic Xtalk is driven by the culprit's load, since this latter dictates the current in the primary loop. When conductors are embedded in a dielectric, like discrete or PCB traces, capacitive Xtalk is increased by the dielectric constant $\varepsilon r = 3\text{–}3.5$ for PVC, 4.5 for epoxy. Magnetic Xtalk is not affected by the dielectric.

Unaware designers tend to regard Crosstalk as a minor "oh-by-the-way" nuisance that will be taken care of, if it happens. The following numerical examples give some measure of this insidious coupling that can cause digital upset and eat-up the EMC margin. It can also export internal circuits' noise to external cables that may in turn violate radiated EMI limits (Table 4.1).

Numerical Example 4.1
Crosstalk in a wire bundle.

Cable bundle: two wire pairs, untwisted/unshielded, in PVC coating, running parallel over 3 m.

Separation s = 3 mm, Height h = 5 mm, wires diam. = 0.5 mm.

Victim impedance: 100 Ω Culprit pair load impedance: 75 Ω.

Estimate Capacitive and Magnetic Crosstalk if culprit signal is **12 V** with **100 ns risetime**.

From curves Fig. 4.3: $C_{1\text{-}2} \approx 6$ pF/m.

$$\text{Correcting for length and dielectric} : C_{1\text{-}2} \approx 6\,\text{pF} / \text{m} \times 3\ \text{m} \times 3.5 \approx 60\,\text{pF}$$

$$M_{1\text{-}2} = 0.2\,\mu H / \text{m} \times 3\,\text{m} = 0.6\,\mu H \left(\text{there is no dielectric effect for } M_{1\text{-}2}\right)$$

Table 4.1 Examples of crosstalk

	Freq. (MHz)	1	3	10	30	50
Two wire pairs, AWG24, separated by 3 mm PVC insulation, parallel length 2 m	**Cap. Xtalk (dB)**	−50	−40	−30	−20	−16
Culprit and victim terminated in 100 Ω	**Magn. Xtalk (dB)**	−30 (3%)	−20 (10%)	−10 (30%)	−4 (60%)	(asymptote)
Two traces on Multiayer PCB	**Culprit circ. rise time tr:**	100 ns	30 ns	10 ns	3 ns	
Edge-to-edge trace separ. 0.3 mm Culprit and victim terminated in 120 Ω						
Parallel length 10 cm	**Cap. Xtalk (dB)**	−46	−36	−26	−16	
	Victim's voltage for 1 V culprit. voltage	(5 mV)	(15 mV)	(50 mV)	(150 mV)	

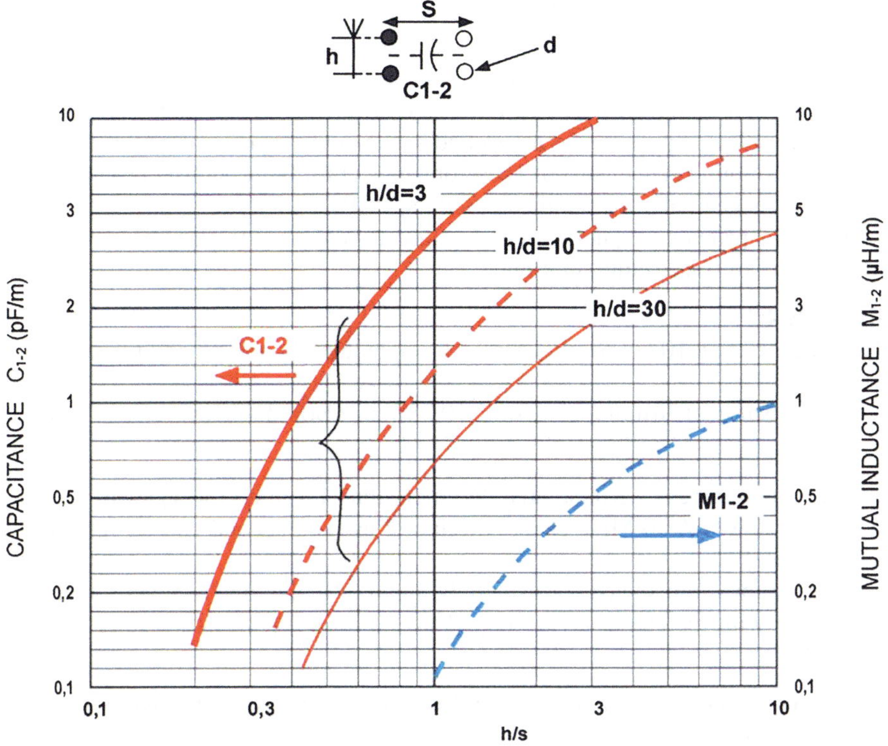

Fig. 4.3 Coupling capacitance (C_{1-2}) and mutual inductance (M_{1-2}) between two wire pairs C_{1-2} is given in air. If wires are embedded in a dielectric, multiply C_{1-2} by $\varepsilon r = 3.5$ (for PVC), or $\varepsilon r = 2$ (for Teflon). C_{1-2} is influenced by the wire diameter, while inductance (M_{1-2}) is not

From Eq. (4.3), *for capacitive Xtalk*:

$$\text{Vvict}\left(\text{cap}\right) = 12\text{V} \times \text{R}_v\, \text{C}_{1-2} \,/\, \Delta t$$
$$= 12 \times 100 \times 60 \cdot 10^{-12} \,/\, 100 \cdot 10^{-9} = \mathbf{0.7\,V}$$

From Eq. (4.4), for magnetic Xtalk:

$$\text{Vvict}\left(\text{mag}\right) = 12\text{V} \times \left(\text{M}_{1-2} \,/\, \text{R}_{\text{culp}}\right) / \Delta t$$
$$= 12\text{V} \times \left(0.6 \cdot 10^{-6} \,/\, 75\right) / 100 \cdot 10^{-9} \approx \mathbf{1\ V}$$

The two effects are in the range of a volt. Depending on which end of the pair is the highest victim resistance, Capacitive and Magnetic Xtalk may add-up (Near-end Xtalk) or subtract (Far-end Xtalk). Assuming the worst, we could have up to 1.7 V appearing at the victim input.

4.1.1 Crosstalk in PCBs

PCB Xtalk is merely a variation of wire-to-wire coupling, the round conductors being replaced by flat stripes. One important difference is that in epoxy, the trace-to-trace coupling capacitance is multiplied by 4.5 compared to ordinary wires, while magnetic coupling is unchanged. Therefore, except for large culprit currents, capacitive Xtalk is generally the dominant mode in PCBs.

Table 4.2 provides a quick tool giving directly the trace-to-trace Xtalk in dB/cm for a variety of traces configurations.

Example 4.2
Two traces have a 10 cm parallel run on a single layer board, no gnd plane, with:

- $w = 15\,\text{mils}\,(0.38\,\text{mm}) \cdot s = 15\,\text{mils}\,(0.38\,\text{mm}) \cdot$ Board thickness, $h = 1.2\,\text{mm}$

- All ground traces or ground areas are more than 1.2 mm (3 × w) away.
- Culprit is a 5 V, 30 MHz clock with tr = 2 ns (i.e., second corner frequency is 160 MHz).
- The Victim's impedance consists in the input of a data line receiver ($\gg$1 kΩ) in parallel with the characteristic impedance of an I/O pair, $Z_0 = 130\ \Omega$, so $Z_{\text{victim}} \approx 130\ \Omega$.

Estimate Xtalk at harmonic #3, 5, 9 frequencies with respect to RF pollution of the nearby I/O wires.

Solution

W/h ratio is: 0.38 mm/1.2 mm = 0.3 and there is no ground plane underneath (Table right-hand column). s/h ratio: 0.38/1.2 = 0.3.

Harmonic	#3	#5	#9
Frequency:	90 MHz	150 MHz	270 MHz
Xtalk from table:	−30 dB	26 dB	−21 dB
(corresponding ratio)	3%	5%	9%
Correction for Z_{victim}:			
130 Ω/100 Ω: 1.3 **Xtalk, final:**	**4%**	**6.5%**	**11.7%**

Table 4.2 Capacitive Crosstalk between PCB traces, in dB, for R_v(total) = 100 Ω and 1 cm length. For other values of length and R_v, apply the correction: 20 log [1 (cm)·R_v/100]. The clamp in dB on the bottom line is the maximum possible crosstalk, ever. For buried traces, this high frequency clamp is 10 dB lower

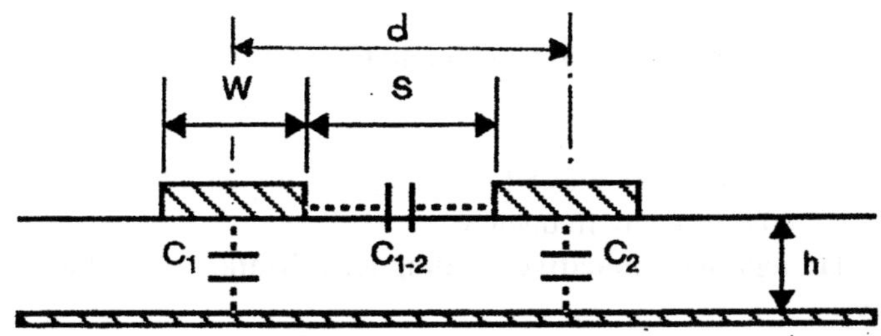

C_2 Z_0	w/h=3 1.6pF/cm 40Ω			w/h=1 0.8pF/cm 70Ω				w/h=0.3 0.5pF/cm 120Ω				w̃/h=0.3-0.6 No Gnd Plane			
s/h	10	3	1	10	3	1	0.3	10	3	1	0.3	10	3	1	0.3
corresp. d/h	(13)	(6)	(4)	(11)	(4)	(2)	(1.3)	(10.3)	(3.3)	(1.3)	(.6)	(10.4)	(3.4)	(1.4)	(.9)
C_{1-2} (pF/cm)	.016	.06	.16	.007	.04	.14	.25	.003	.03	.1	.2	.18	.22	.35	.5
Freq.															
1kHz	-160	-148	-140	-167	-152	-141	-136	-174	-154	-144	-138	-139	-137	-133	-130
3kHz	-150	-138	-130	-157	-142	-131	-126	-164	-144	-134	-128	-129	-127	-123	-120
10kHz	-140	-128	-120	-147	-132	-121	-116	-154	-134	-124	-118	-119	-117	-113	-110
30kHz	-130	-118	-110	-137	-122	-111	-106	-144	-124	-114	-108	-109	-107	-103	-100
100kHz	-120	-108	-100	-127	-112	-101	-96	-134	-114	-104	-98	-99	-97	-93	-90
300kHz	-110	-98	-90	-117	-102	-91	-86	-124	-104	-94	-88	-89	-87	-83	-80
1MHz	-100	-88	-80	-107	-92	-81	-76	-114	-94	-84	-78	-79	-77	-73	-70
3MHz	-90	-78	-70	-97	-82	-71	-66	-104	-84	-74	-68	-69	-67	-63	-60
10MHz	-80	-68	-60	-87	-72	-61	-56	-94	-74	-64	-58	-59	-57	-53	-50
30MHz	-70	-58	-50	-77	-62	-51	-46	-84	-64	-54	-48	-49	-47	-43	-40
100MHz	-60	-48	-40	-67	-52	-41	-36	-74	-54	-44	-38	-39	-37	-33	-30
300MHz	-50	-38	-30	-57	-42	-31	-26	-64	-44	-34	-28	-29	-27	-23	-20
1GHz	-43	-31	-23	-47	-32	-21	-18	-54	-34	-24	-18	-19	-17	-13	-10
Asymp.	-40	-29	-21	-40	-26	-16	-12	-44	-25	-16	-11	-7	-6	-4	-3

$$\text{Crosstalk} = 20 \log \frac{R_{victim} C_{1-2} \omega}{\sqrt{[R_V \omega (C_2 + C_{1-2})]^2 + 1}}$$

4.2   Solutions Against Crosstalk

All solutions that work against field-to-cable coupling are efficient against Crosstalk. Furthermore, besides Rules #6, 7, 8, 9 seen before, a few specific solutions can be applied:

Rule #10: against capacitive or magnetic Xtalk, increase the culprit-victim wires separation.

Try to increase the s/h ratio, since mutual capacitance and inductance fall-off rapidly when s/h ratio becomes >1. A s/h $\geq$10 is a guarantee that Xtalk will never exceed 1% (−40 dB) at any frequency.

Rule #11: against capacitive or magnetic Xtalk, twist the victim, or culprit, wires.

Make sure to twist together the + and − wires. Twisting on the culprit pair maybe the most efficient because this can protect several victims' pairs in a same bundle.

Rule #12: against Xtalk in PCBs, use the « poor man's » shield, a simple ground trace between culprit and victim traces.

This simple, unexpensive precaution easily reduces both Xtalks by 14–20 dB (a 5–10 factor).

All the coupling mechanisms seen above and their reduction techniques are perfectly reciprocal: they will prevent radiated emissions as well.

Quiz

There is ONE good answer, or ONE that is better than the others.

1. To reduce Capacitive Crosstalk between two wires

   (a) decrease the wire-to-ground capacitance, by bringing the wires closer to their ground (or return)
   (b) increase the wire-to-ground capacitance, by bringing the wires closer to their ground (or return)
   (c) increase the wire-to-wire capacitance
   (d) decrease the wire-to-wire capacitance, and increase the wire-to-ground capacitance

2. Capacitive Crosstalk increase when:

   (a) victim impedance is high
   (b) both culprit and victim impedances are high
   (c) culprit current is higher
   (d) one end of the victim circuit is shorted to ground

3. Both Capacitive and Magnetic Xtalk are frequency dependent, but:

   (a) capacitive Xtalk increase like $(F)^2$ while Magnetic Xtalk increase like F
   (b) capacitive Xtalk saturates more rapidly when frequency increase
   (c) both capacitive and magnetic Xtalk increase with frequency
   (d) both capacitive and magnetic Xtalk increase with frequency until they approach 0 dB

4. When estimating a Crosstalk coefficient from a 5 V culprit voltage:

    (a) 0 dB Xtalk means that there is no voltage appearing across victim input
    (b) −46 dB Xtalk means 25 mV appearing across victim input
    (c) −46 dB Xtalk means 2 mV appearing across victim input
    (d) −46 dB Xtalk means 200 mV appearing across victim input

5. Against Crosstalk,

    (a) twisting the culprit wires together reduce only magnetic crosstalk
    (b) twisting the culprit wires together reduce both capacitive and magnetic crosstalk
    (c) twist each culprit wire with the closest victim wire to achieve better flux cancelation
    (d) twisting is useless against Xtalk since it is a Common Mode type of coupling

Chapter 5
Shielded Cables: Their Role in Reducing EMI Susceptibility and Emissions

5.1 Introduction

Before addressing the last coupling path, that is, the coupling occurring from (or to) the power mains, it is time to review a solution that is widely involved in controlling conducted, radiated, and crosstalk EMI situations: the use of **shielded conductors**.

Although anyone can intuitively foresee that putting wire (or wires) inside a shield will protect them from interference, the subject is not that simple and requires some deeper explanation. This chapter will explain as clearly as possible for the non-specialist why and how a cable shield works, and how much EMI reduction can be expected. It will also explain why the choice of certain cables or installation practices will result in disappointing results.

5.2 Basic Role of a Shield over a Cable Link

As soon as an equipment is fitted with external cables whose length exceeds the largest box dimension, it is highly probable that these will be the largest contributors to radiated susceptibility and emissions, at least up to several hundred Megahertz. To some extent, cables are also involved in the common impedance coupling, conducted path.

Although shielding a cable may appear as the obvious, solid barrier to radiated coupling to or from the wiring inside, application may not be so easy. Throwing-in shielded cables at the last minute may give disappointing or even disastrous results. The author has even seen odd cases where shielded cables increased the radiated EMI levels at some frequencies. There are explanations to this, of course, as will be seen.

© The Author(s), under exclusive license to Springer Nature Switzerland AG 2025
M. Mardiguian, *ElectroMagnetic Compatibility*,
https://doi.org/10.1007/978-3-032-02688-0_5

The basic principle for a shield to work against all types of EMI, with the widest coverage of situations (E-field, H-field, low and high frequency, Diff. Mode/Comm. Mode, etc.), is to create a continuous barrier enclosing the conductors and 360° bonded to the conductive boxes at both ends. No matter which theory is applied: reflection loss, absorption loss, Faraday cage effect, mutual inductance, etc. ad infinitum, calculations, and experiments show that when an entire system is enclosed in a continuous barrier, its sensitivity to EMI is reduced. This is true, regardless *this barrier is earthed or not* (Fig. 5.1).

If the boxes are not full metallic envelopes, the principle still can work, provided there is at least one large metal face or ground plane to connect the shield on both ends, closing the cable-to-shield return path for CM currents. Otherwise, as in the case of solid plastic boxes, a cable shield without a reference plate for grounding its ends will not be efficient against radiated susceptibility or emissions. For such case, I/O port decoupling and ferrite loading would be more appropriate if no more than 20–30 dB reduction is needed.

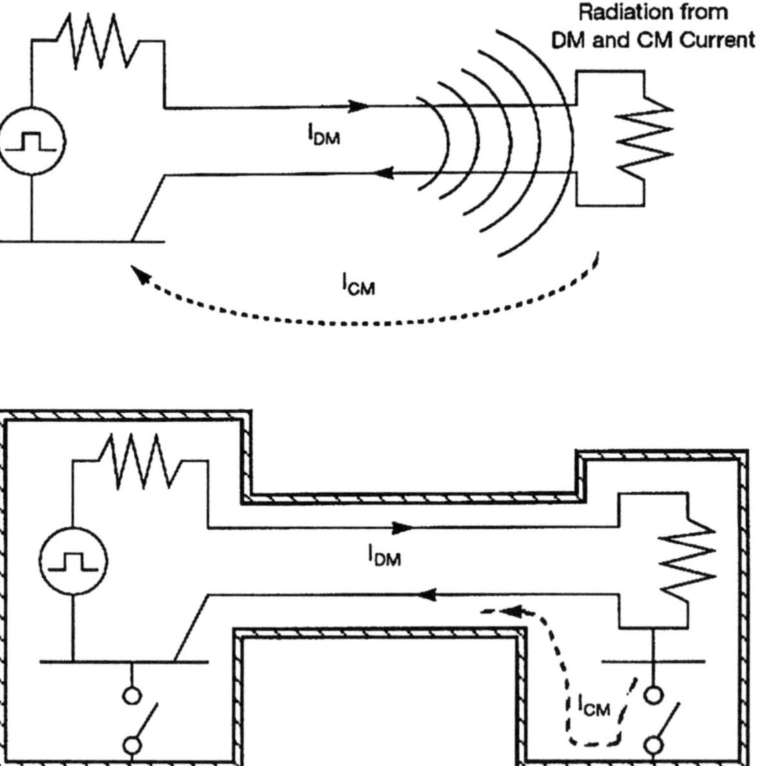

Fig. 5.1 An ideal shield. Provided the metal barrier is uninterrupted and homogeneous, radiation is strongly reduced, *whether or not the inner circuit is grounded to the shield, or the shield connected to earth*

5.3 Principal Types of Cable Shields

Cables shield are seldom solid tubes, welded at each end to the system boxes or subassemblies (exceptions are semi-rigid coax or some specific military systems). Yet, the concept of a conductive sheath surrounding the wires can still be achieved by other constructions. Many technologies are available for cable shields:

Tinned copper braid (single or double layer)
Aluminum foil or aluminized Mylar folded over the cable like cigarette paper
Aluminum foil + copper braid
Thin metallic tissue (silver, stainless steel, copper)
Stretched metal foil
Tinned copper or tinned steel spiral wrap
High permeability wrap, associated with one or several layers of copper braid
Corrugated, bellow-style cable shield
Semi-rigid copper shield (essentially used for some RF coaxial links)

What are their respective merits? How much attenuation can we expect, and in what applications? How much is enough? What is the impact of the shield termination hardware at the equipment barrier, and can it be predicted? These will now be explained (Fig. 5.2).

5.3.1 The Two Basic Families of Shielded Cables

Although any conductor(s) slipped in a metallic sheet can be labeled as "shielded," there are two basic types of shielded cables: coaxial cables and shielded pairs or multipairs. Both types reduce the interference received or generated by the active conductors: HF ground loop coupling, Xtalk, field induction, but they present a fundamental difference (Fig. 5.3).

5.3.2 Coaxial Cables

In a coaxial cable, the shield is altogether:

– the return path for the intended signal,
– an alternate, preferred path for the undesired noise current, whatever it is received (Susceptibility) or generated (Emission) by the system.

This carries a specific constraint for a coaxial link, expressed in the following rules:

RS-1: *With a coaxial cable, the shield must be connected to the signal reference at both ends, for functional reasons, and to the equipment chassis, for EMC reasons.*

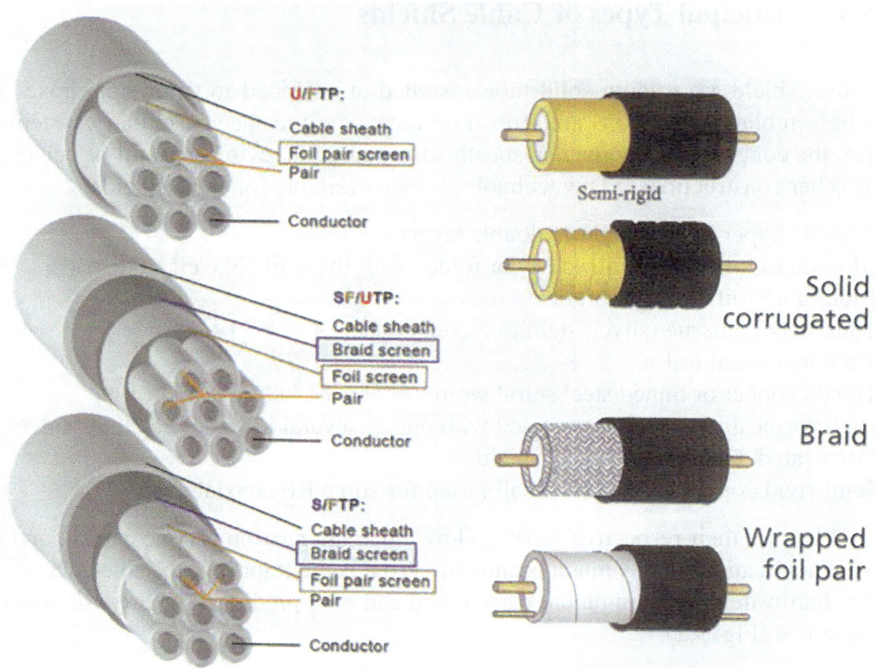

Fig. 5.2 Various constructions of cable shields. Left: shielded pairs, right: coaxial cables

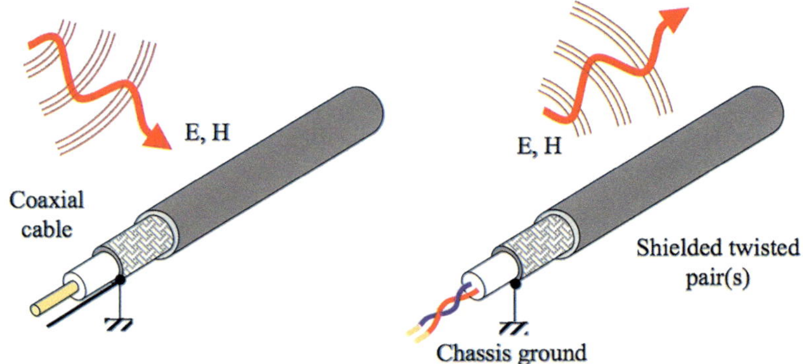

Fig. 5.3 The two families of shielded cables. Notice (left) that for a coaxial the shield need to be tied to the electronic "ground" for functional reasons, and to the chassis for best EMC efficiency

Thus, the normal termination of a coaxial cable is forcing the 0 V to be grounded to the equipment chassis through the I/O port. Although this is generally the recommended configuration against high frequency EMI (see multi-point ground, Chap. 3), there are cases where the designer has opted for a single point (or star) ground arrangement. In such cases, an *additional rule should be followed*.

RS-2: *If, functionally, signal interface requires an isolated 0 V, a coaxial link is not the best choice, unless a galvanic isolation device (signal transformer or opto-isolator) is used.*

Some equipment designers, sticking to the Single Point Gnd rule (SGP) keep the coaxial shield grounded to the 0 V, but floated from the chassis in order to prevent a ground loop. Although this opens the loop at low frequency, it turns out as a disaster in case of high frequency EMI: the coaxial shield will collect the EMI currents and dump them onto the signal reference, which is usually a critical conductor. Figure 5.4 presents a trade-off to this dilemma. Closest to the I/O port, the shield is connected to the chassis via a "zero-inductance," leadless ceramic capacitor of low value, generally a few nanoFarads.

A few specific advantages beneficial to EMC performance can be credited to the coaxial cable:

- Thanks to tight manufacturing tolerances, parameters like low HF line losses and characteristic impedance are guaranteed with a good accuracy.
- Since the coaxial cable (invented around the 1920s) has a long history of intensive use in RF engineering and instrumentation, a large inventory of good quality connectors (BNC, N, SMA, etc.) is available, their shielding factor ranging from good to excellent.

5.3.3 Shielded Twisted Pairs and Multiconductors Shielded Cables

Shielded Pairs exhibit a noticeable difference: the shield is no longer an active return conductor. Against susceptibility it is there to neutralize the EMI ambient currents instead of letting them flow in the protected wire pair. Reciprocally, for preventing the signals carried by the pair from circulating in the external cable-to-ground loop, and eventually radiate excessive EMI, the shield will collect these

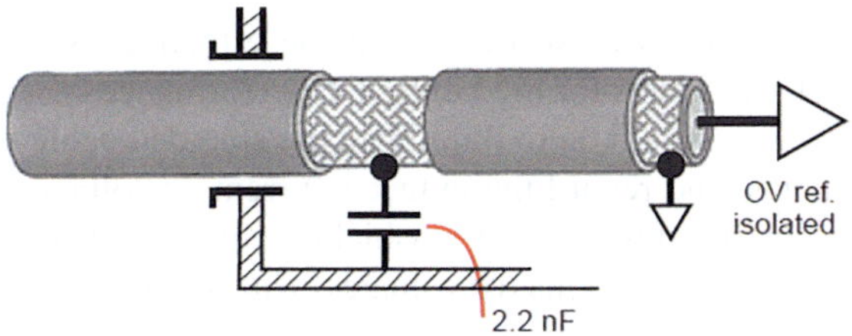

Fig. 5.4 A trade-off for accommodating a coaxial cable port when the 0 V signal reference needs to be isolated from chassis for safety reason or to prevent low-frequency ground loops. The 0 V is floating from chassis at low frequencies (2 nF = 500 kΩ at 150 Hz), but virtually grounded above a few Megahertz

Common Mode currents escaping from the pair, offering them a low-impedance return path back to their source.

Since practically all discrete signal pairs are twisted—an efficient way for preventing field-to-cable Diff. Mode pick-up and Crosstalk, shielded twisted pairs are usually designated as STP, by contrast to unshielded ones (UTP).

Advantage of the STP

Given that the signal current is flowing back and forth in the two wires of the pair, the shield plays no role in returning the intended signal. The designer has all freedom to ground the shield of the STP to the equipment frame, and still keep his 0 V reference isolated, if he so wishes.

Specific STP Disadvantages

Due to the twisting, the accuracy of the cores-to-shield distance is not as perfect as with a coaxial cable, causing more line losses and impairments due to the fluctuating characteristic impedance.

For reasons related with the above, the symmetry of the two wires of the pair vs. ground (that is the shield) is not perfect, causing some % of the signal current to flow in the shield, hence in the system ground, generating EMI emissions. The same is true for susceptibility.

5.4 Evaluating the Merits of a Shielded Cable

For long, shielded cables were used more or less casually, assuming that if a cable link is shielded, it would no longer be a cause of EMI concern. In fact, like any element of a system, the global quality of a shielded link must be quantified. This includes not only the sole shield performance, but also its terminations, i.e. the connector/receptacle assembly.

Measurement techniques exist, evaluating the effectiveness of a cable shield, along with calculation models for predicting the performance of a given cable, once installed. Simplest way would be to illuminate the shielded cable with a given field at several frequencies, and record the induced current or voltage on the inner conductor, then to repeat the test with an unshielded version of the same cable. The comparison of the induced currents (or voltages) with/without the shield would give a figure of the shield performance.

Unfortunately, this is an expensive test, requiring sets of antennas and an anechoic room, bearing the uncertainty inherent to any radiated measurement. Furthermore, the results for a same cable sample would vary depending on the type of radiating antenna used in the test (H-field loop, Dipole, etc.), the near-field or far-field conditions of the set-up, and the height above ground for the tested item. A better method consists in measuring the shield transfer impedance, Zt, as explained next.

5.4.1 Shield Transfer Impedance, Zt

A convenient way of characterizing the merit of a cable shield [2, 3, 5, 6] is its transfer impedance, Zt. It relates the current flowing on a shield surface to the *longitudinal voltage* it develops on the other side of this surface (Fig. 5.5). This voltage is due to a diffusion current through the shield thickness (with a solid tube, this diffusion rapidly becomes unmeasurable at high frequency, due to skin effect), and to the leakage inductance through the braid's holes. The better the quality of the braid, the less the longitudinal shield's voltage.

Let us start with the simple configuration of a coaxial cable exposed to an EMI threat. As a result, an undesirable current is flowing along the shield. Since the

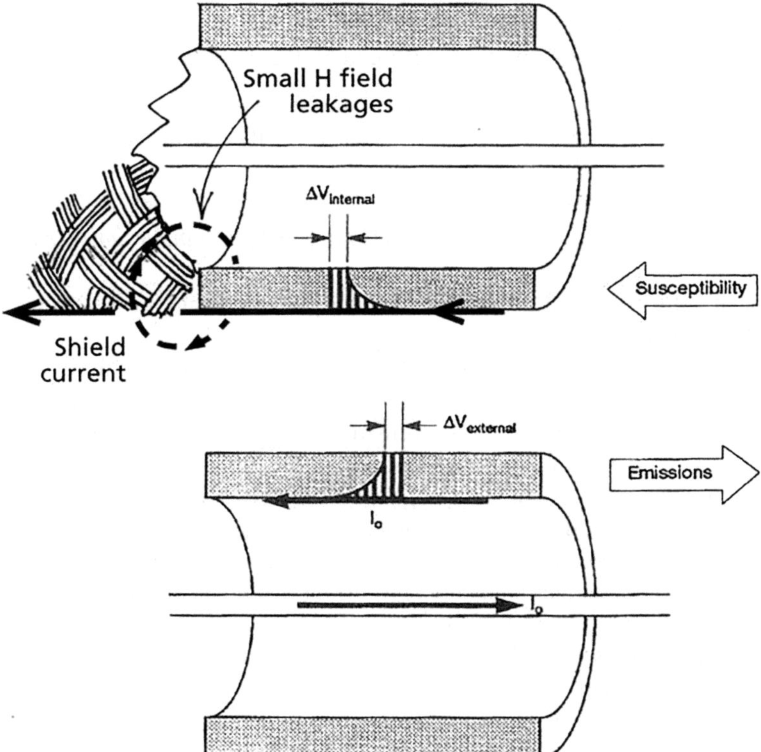

Fig. 5.5 Conceptual view of the transfer impedance Zt, showing both the ohmic resistive and the magnetic field leakage effects

shield wall is not a perfect tubular conductor, the flow of current is encountering two mechanisms:

(a) *The braid dc resistance* (typ. 5–20 mΩ per meter for a single braid). Figure 5.5 shows the current density decreasing progressively in the thickness of the shield, such as when frequency increases, the skin effect tends to concentrate more current on the shield surface that is looking toward the source, leaving less and less current on the opposite surface.

(b) *The maze of small leakages* caused by the holes in the braid weaving. This is defined as a leakage inductance in nH/m. As a result, a small voltage appears along the internal core-to-shield space.

If used for susceptibility calculations and normalized to a 1 m sample, Zt is defined as:

$$Zt\left(\Omega/m\right) = Vi/\left(I_{sh}\right) \tag{5.1}$$

where

Vi = longitudinal voltage induced inside the shield, causing a noise current to circulate in the center conductor.

I_{sh} = current injected artificially on the shield by the external EMI source

The term Zt itself include the shield resistance R_{sh} and the shield transfer inductance L_t, regarded as the leakage inductance from the inside-out (or the reverse), such as:

$$Zt\left(\Omega/m\right) = R_{sh}\left(\Omega/m\right) + j\omega L_t\left(H/m\right) \quad (2)^* reminder : \omega = 2\pi F \tag{5.2}$$

Typical values of R_{sh}, Lt for a decent quality, single braid are 10–15 mΩ/m and 1–2 nH/m respectively.

The above is a gross approximation. The actual mechanism is more complex: in fact it is the mutual inductance between the shield and the inner conductor that plays a major role. If we call L_1, the loop formed by the center conductor alone and the ground, and L_2 the shield-to-ground loop, there is a strong mutual inductance M_{1-2} between these two loops. Due to a tight coupling between the inner conductor and its surrounding shield, this mutual inductance is almost equal to the self-inductance of the center conductor vs. ground. The result is that the current in the shield will induce in the inner conductor an opposite current that tends to cancel the initial EMI current.

The cancelation is never 100% but can come pretty close. With a good quality single braid, it reaches 99.7%, that is only 0.3% of the initial noise current remains in the inner wire. This canceling effect leads to a third rule, essential to the functioning of a shielded cable:

Rule S-3: *For its good operation, a cable shield must carry a current equal and opposite to the total, net, current carried by the inner conductor. Therefore it must be grounded (not necessarily earthed) to the equipment boxes at both ends.*

This is essential for the shield to work. A good test is: with a perfect shield, *a current probe clamped around the whole cable should read NO current,* meaning that the net current flowing in the inner conductor is perfectly balanced by an equal, opposite current in the shield.

Typical values of Zt for various cables are shown in Fig. 5.6 . If the shield is grounded by pigtails (a poor practice), pigtails and other impedances must be added to the Zt and loop impedances calculations.

Once the external EMI current in the loop (I_{sh}) is known (measured or calculated), the noise voltage induced internally can be derived for any length of this cable by:[1]

$$Vi = Zt(\Omega/m) \times (I_{sh}) \times l(m) \qquad (5.3)$$

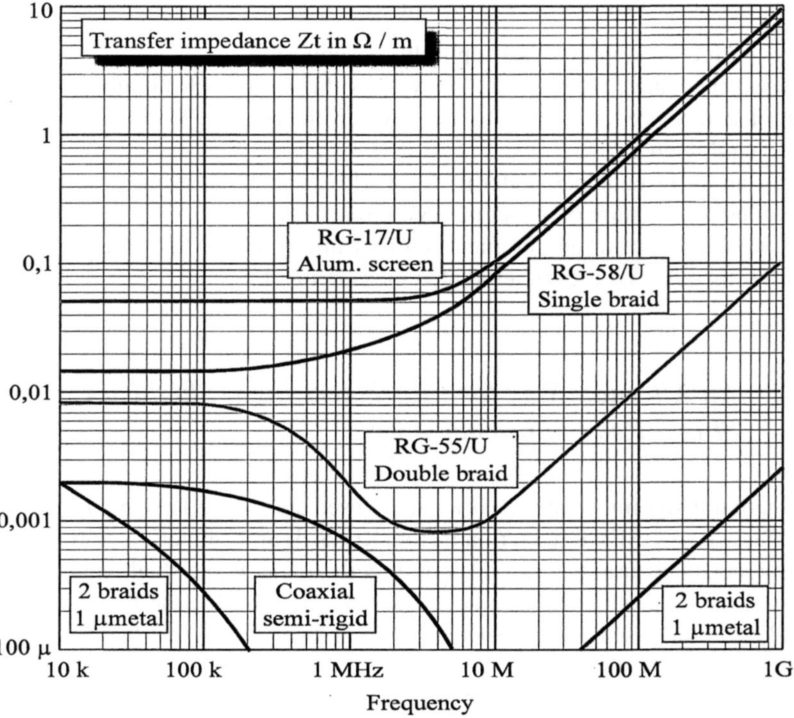

Fig. 5.6 Typ. values of Zt for a 1 m sample. Above 150 MHz, since $l > \lambda/2$, values are indicative only

---

[1] Caution: One must check that the physical length of the exposed cable does not exceed $\lambda/2$.

If $l \geq \lambda/2$, the voltage calculation must be limited to an half-wave length, that is $l \approx 150/F(MHz)$.

The principle is perfectly reciprocal and applied to emissions as well. For the 1 m sample, the internal signal current I_0 returning by the shield's inner surface causes an EMI voltage to appear along the outer side of the shield, which can be expressed as:

$$V_{ext} = I_0 \times Zt\left(\Omega/m\right) \tag{5.4}$$

$$= Zt\left(\Omega/m\right) \times V_0/Z_L \tag{5.4bis}$$

where

V_{ext} = external voltage appearing along the shield-to-ground loop
V_0, I_0 = signal voltage and current on load side (Z_L) of the coaxial cable

This voltage, in turn, excites the antenna formed by the external cable-to-ground loop.

The key advantages of the Zt concept are:

– It is perfectly reciprocal (susceptibility $\leftrightarrow$ emission).
– It does not matter if the shield current is due to a field illuminating the cable (Radiated Susceptibility) or to a conducted interference (Ground shift between two equipment).
– Zt is an intrinseque parameter to the shielded cable, independent of the radiated or conducted nature of an actual EMI threat.
– Being a conducted measurement (current injection over the shield), it does not suffer the uncertainties of a radiated measurement.

5.4.2 An Alternate Way for Characterizing a Shielded Cable: Shield Reduction Factor Kr

Although transfer impedance Zt is a widely used and dependable parameter, shielding effectiveness (SE) or reduction factor (Kr) as figures of merit are often preferred by designers, because they can relate it directly to the whole shielding performance required for the system. Requiring 60 dB of shielding for the boxes of a system would be a nonsense if the associated cables and connecting hardware provide only 20 dB, and vice versa.

Practical formulas, directly expressing the shielding factor Kr of a cable, given its Zt (Ω/m) have been devised [2]. This shielding factor *Kr becomes a dimensionless number in dB* that incorporates Zt, but allows for a direct prediction for an installed shielded cable.

Regarding susceptibility, Shield reduction factor (Kr) is the ratio of the differential mode voltage (V_{diff}) appearing, core-to shield at the receiving end of the cable, to the external common mode voltage (V_{cm}) applied in series into the loop. It can be expressed by:

$$Kr(dB) = 20\log\left(V_{diff} / V_{cm}\right) \tag{5.5}$$

Regarding emission, a reciprocal definition, similar to the basis of Eq. (5.4) can be used for characterizing a shielded cable with respect to emission, simply by the ratio of the Common Mode Voltage (V_{cm}) appearing in series into the external loop, to the differential mode signal voltage (V_d) applied, core-to shield at one end of the cable.

$$Kr(dB) = 20\log\left(V_{cm}\,\text{external} / V_{diff}\,\text{internal}\right) \tag{5.5bis}$$

Calculations and experiments have shown that the Kr factor is the same in the two above cases. Kr could also be regarded as the Mode Conversion Ratio between the internal circuit (center conductor and shield) and the external one (the shield-to-ground line). One could also compare the current in the loop if the shield was not there, to the remaining inner circuit current when the shield is in place, grounded both ends (Fig. 5.7).

A complete demonstration leading to the expression of Kr can be found in [2]. We will just give the end results:

$$Kr = \frac{R_{sh} + j\omega L_t}{R_{sh} + j\omega L_{ext}} \tag{5.6}$$

where

R_{sh} = shield resistance in Ω/m
L_{ext} = self-inductance of the external shield-to-ground loop
L_t = Transfer inductance of shield[2]

This expression unveils three frequency domains,

(a) **Very Low Frequencies**: the term ωL_t is negligible, Zt is dominated by resistance R_{sh}.
   Kr = $R_{sh}/(R_{sh} + j\omega\,L_{ext}) \approx 1\,(0\,dB)$ below few kilohertz, since the lower term, loop impedance reduces to R_{sh}
(b) **Medium Frequencies** (typically above 5–10 kHz for ordinary braided shield):

$$Kr = \left(R_{sh} + j\omega L_t\right) / \left(j\omega L_{ext}\right)$$

Here the Reduction Factor increases linearly with frequency

---

[2] While L_t is intrinsic to the cable shield alone, L_{ext} depends on the height of the cable above ground. Therefore the later is installation-dependent.

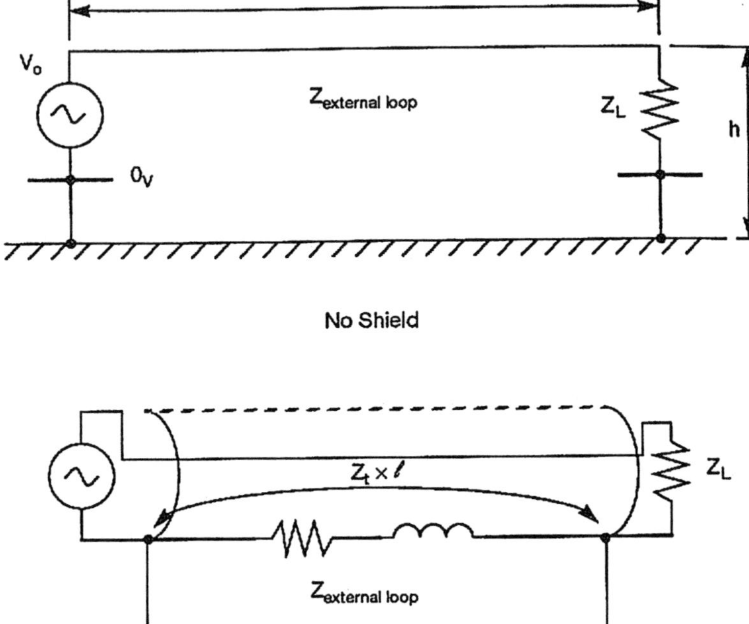

Fig. 5.7 Simplified view of the shield reduction factor (Kr) definition. It compares the currents in the internal circuit, with and without the shield in place

(c) **Higher frequencies** (typically above 1 MHz, up to first $\lambda/2$ resonance), reduction factor stays constant, *independent of length and frequency*:

$$Kr = L_t / L_{ext}$$

A general, handy formula is derived, valid for any frequency *from 10 kHz up to first <$\lambda/2$ resonance*. It takes into account a mean value for the height/diam value of the cable above ground.

$$Kr(dB) = -20\log\left[1 + 6F((MHz))/Zt(\Omega/m)\right] \tag{5.7}$$

The value entered for Zt must be the one taken at the frequency of concern

5.4.3   *Kr Values When Cable Length Is Approaching or Exceeding λ/2*

As already mentioned for Zt, when the cable dimension reaches a half-wave length, the shield is no longer carrying a uniform current. The "electrically short line" assumption becomes progressively less and less acceptable when cable length l exceeds λ/10. The shield grounded both ends behaves as a dipole exhibiting self-resonance and anti-resonance for every odd and even multiple of λ/2, respectively.

Actual wave propagation in the loop being 0.7 times slower than in free space, the effective wavelength is recalculated for the actual resonances. At these frequencies, the shield current will exhibit peaks, resulting in ≈10 dB periodic degradations of Kr factor. This reflects the actual situation where, for a uniform EMI stimulus, the resulting interference will show periodic "humps" beyond the first resonance point.

Taking typical values for the shield-to-ground characteristic impedance with a conservative approach aligned on the asymptote of the humps, we reach a simple expression for **worst case Kr** beyond the first resonance:

Both *Susceptibility and Emission* cases, above λ/2 resonance:

$$\text{Kr}_{(\text{min})}\left(\text{dB}\right) = -20\log\left[210 / 0.7 L_t \left(\text{nH} / \text{m}\right)\right]$$
$$= -20\log\left[300 / L_t \left(\text{nH} / \text{m}\right)\right] \tag{5.8}$$

As a recap of *Kr for below and above resonance conditions*:

For cable length <λ/2: Kr (dB) = -20 Log[1 + (6F(MHz))/Zt(Ω/m)]

For cable length >λ/2: Kr min (dB) = -20 Log[300/L_t(nH/m)]

Numerical Example
An RG-174 coax is connecting two racks.
   Useful signal: 15 MHz analog video, with 3 mV detection sensitivity.
   The cable parameters are:

Length: 2 m, Average height above the metallic frame = 10 cm.
Good quality coaxial connectors are used at both ends (2.5 mΩ/connector).

When exposed to an undesirable 30 MHz, 50 V/m field, a Comm. Mode voltage of 6 V is induced in the shield-to-frame loop (see Chap. 4 "Radiated coupling"). What is the voltage induced internally?

Solution

For 30 MHz, cable length is $<\lambda/2$. We can use Eq. (5.7) for Kr. Curve Fig. 5.6 indicates 0.3 Ω/m for the RG174 at 30 MHz. The contact impedance of the good quality connectors (0.005 Ω) can be neglected.

$$Kr\,(dB) = -20\log\left[1 + \left(6F(MHz)\right)/Zt\,(\Omega/m)\right]$$
$$= -20\log\left(1 + 6\times 30/0.3\right) = -55dB \text{ corresponding to } \approx 1/600 \text{ reduction factor}$$

The voltage on the center conductor will be 6 V/600 = 10 mV. This is 3.3 times above the threshold of video sensitivity. We do not know the rejection of the base-band video amplifier for a 30 MHz signal.

Several solutions can reduce the coupled by a three times factor:

1. Select a coaxial cable with a lower Zt, that is, Zt $\leq$0.1 Ω/m at 30 MHz. This is achievable with optimized braided shield (thicker, denser braid) or more easily, with more costly double-braid shield.
2. Slip a large ferrite bead over the cable shield. It will take an added series impedance of about 1000 Ω to achieve the required attenuation. Passing the cable three times into a large bead will provide such impedance.
3. Decrease cable height above chassis ground by at least three times.

A Few Practical Results for Shield Factor Kr, Below and Above First $\lambda/2$ Resonance

Figure 5.8 shows calculated results for three coaxial cables, 1 m above ground, with good 360° contact at connector socket. Curves are valid for any length, provided that the resonance region is adjusted if length is different from 1 m. Figure 5.9 shows test results for a 5 m coaxial cable with shield intentionally spoiled by a 10 cm pigtail. Deterioration of Kr above 8 MHz is spectacular. Notice that below the MHz region, Kr degrades progressively down to almost 0 dB around a few kHz: a shielded cable has no effect against low-frequency Comm. Mode coupling.

5.4.4   Field Radiated by a Coaxial Cable

Most RF signals, base-band video, some LAN links and other HF signals are carried on coaxial cables. Provided that the shield is correctly tied to the signal ground ref. at both ends, and preferably also to the chassis by the coaxial sockets, only a very little current (typ. 0.3%–0.1%, above a few MHz) will return by paths other than the shield itself (Fig. 5.10). This external current radiates a small electromagnetic field.

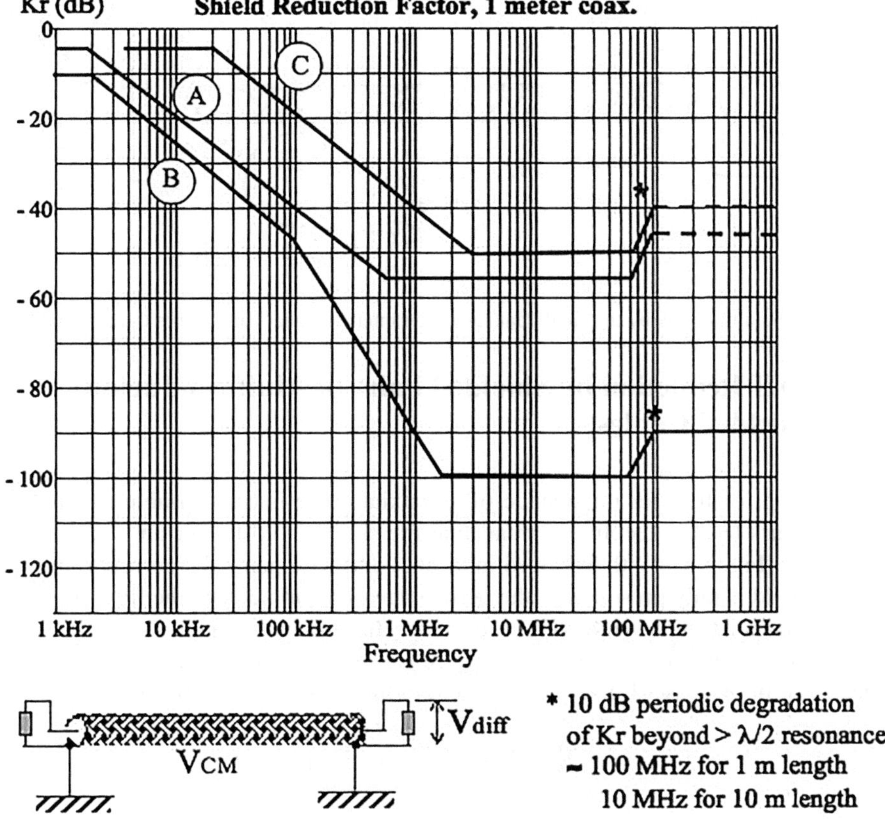

Fig. 5.8 Reduction factor for shielded cables, 1 m above ground: (**A**) RG-58 single-braid coaxial, (**B**) RG-214 Double-braid coaxial (**C**) RG-174 single-braid miniature coaxial, 2 mm outer diameter

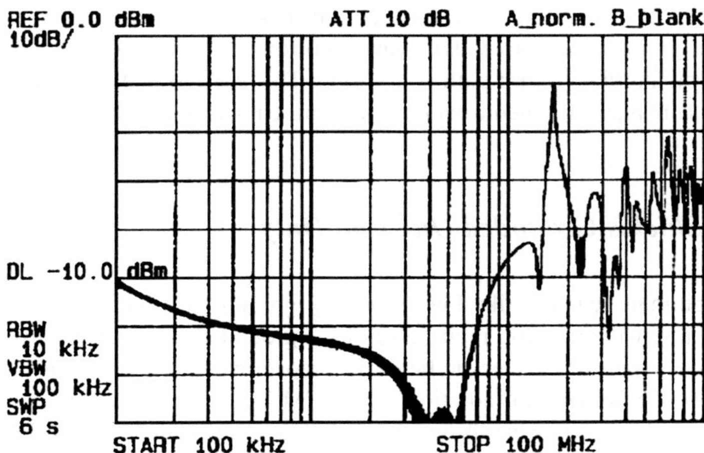

Fig. 5.9 Kr factor for 5 m RG58 cable, shield grounded with 10 cm pigtail. (Courtesy AEMC, France)

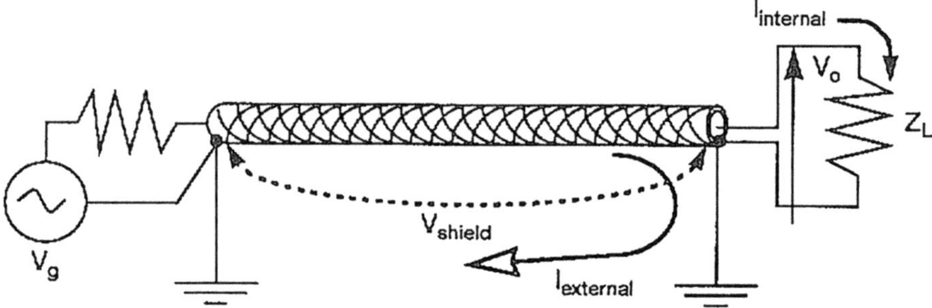

Fig. 5.10 Field radiation by the small leakage of the internal signal current into the external loop

5.5   EMI Reduction by Shielded Pairs or Multiconductor Shielded Cables

The concept of transfer impedance of a coaxial cable is transposable to shielded twisted pairs (STP), keeping in mind that the shield is no longer an active return conductor (Fig. 5.3). With balanced interfaces and wire pairs, the current returning by the shield is only pro-rated to the percentage of asymmetry in the link [3]. If the link is balanced to X% symmetry, the current returning by the shield is, for the worst possible combination of tolerances, only X% of the total current in the loop impedance Z_{EXT}. In this case, Eq. (5.1) applied to radiate susceptibility becomes:

$$V_0 = X\% \left(Zt \left(\Omega / m \right) I_{ext} \right) \tag{5.9}$$

The radiated field is reduced by a factor equal to X%, compared to an ordinary coaxial cable situation. Depending on the quality of the balanced link, X may range from 1% to 10%, a typical (default) value being 5% . Recent progress have been made with high quality (Class #5 STP), with their best balance generally in the 2–3% range. If the STP are interfacing circuits that are not balanced (e.g., the signal references grounded at both ends), a larger portion of the signal current will use the shield as an alternate return path (Fig. 5.11). This portion is difficult to predict: at worst, this unbalanced scheme cannot cause more interference than the coaxial case.

5.5.1   Shields Grounded One End Only

If, for legitimate reasons, like low-frequency ground loops between distant boxes upsetting a sensitive analog input, a shield is grounded at one end only, it will be only effective against LF electric fields and capacitive crosstalk. It has no effect on Comm. Mode immunity or radiation, as the CM loop current does not return by the shield but, rather by the chassis and ground plane, as if there were no shield.

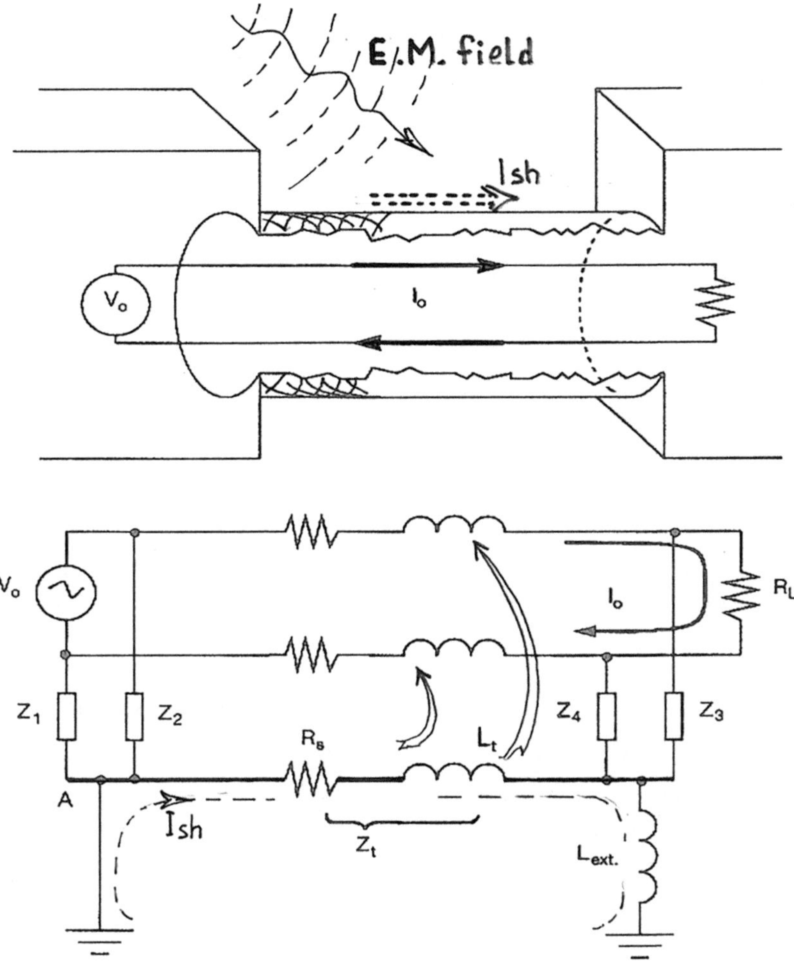

Fig. 5.11 EMI pick-up by a balanced shielded pair

If the frequency of the electric field (or capacitive) threat increases, the capacitive current captured by the STP will increase, and will be flowing on a shield whose impedance also increases (Fig. 5.12). Thus the shield voltage versus ground increases with $(F)^2$, becoming a significant fraction of V_{CM}. The floated end of the shield becomes the "hot" tip of a receiving monopole, and we have just replaced a noisy pair by a noisy shield. So, *a cable shield must be connected at both ends to the boxes, whether these boxes are grounded or not.*

Exceptions are low-level analog instrumentation (strain gages, thermocouples, etc.) and audio interface cables, where only an electrostatic shield is needed. Ground loops are suppressed by galvanic isolation amplifiers, differential amplifiers, and so forth, and grounding a shield at both ends could inject LF (few kHz) noise into the

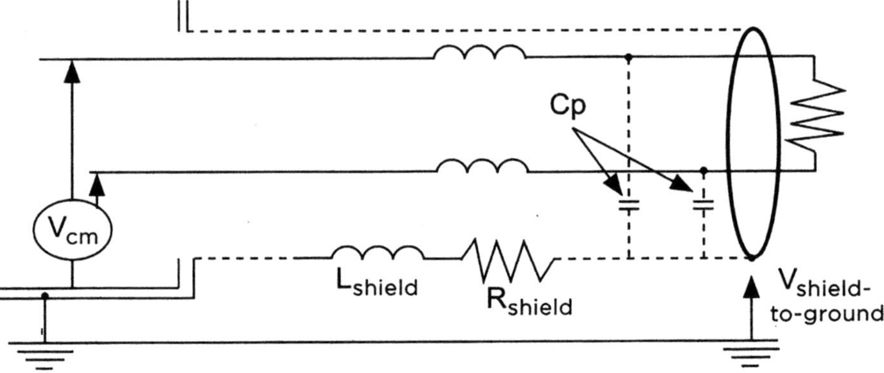

Fig. 5.12 Noise coupling with a floated-end shield

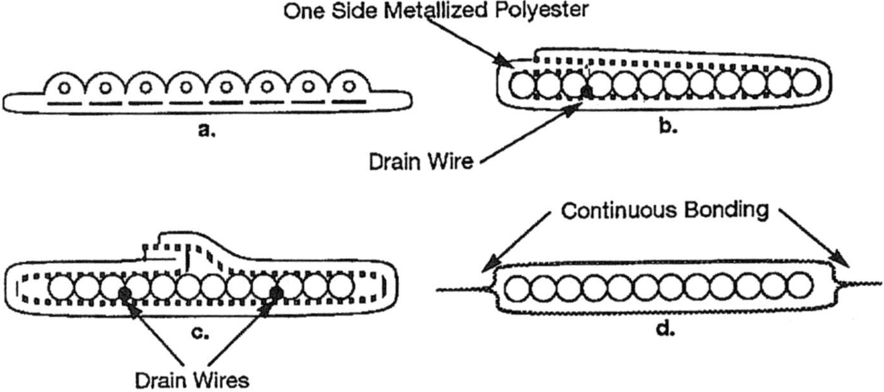

Fig. 5.13 Various shielded flat cable configurations

cable. A few millivolts injected this way are harmless for digital interfaces but can constitute strong interference for low-level analog signals.

5.5.2   Shielded Flat Cables

One specific case of shielded multiconductor cable is the shielded flat cable. A few typical versions are shown in Fig. 5.13. Version (a), sometimes advertised as "shielded," is merely a flat cable with an embedded ground plane. Although offering some advantages, its reduction is often insufficient because CM current can still flow on the single-side foil edges and radiate [4]. The (b) version, also marketed as "shielded" is leaky at HF due to the long, unclosed seam running over the entire length. The drain wire is acceptable as a low-frequency shield connection, but totally inadequate at HF. Versions (c) and (d) deserve to be called "shielded" as the shield fully encircles the wires. Yet, with (c), there is no access to outer metal surface, 360° bonding is not easily made, relying on the sole drain wire.

5.6   Importance of the Shield Connections

As important as a low Zt shield is its low-impedance termination [1] to the equipment metal boxes. The connection impedance Z_{ct} is directly in the signal current return path, in series with Zt (Fig. 5.14). Therefore, Z_{ct} can increase the coupled voltage for a given shield current. The following values are typical transfer impedances of *one* shield end connection (Table 5.1):

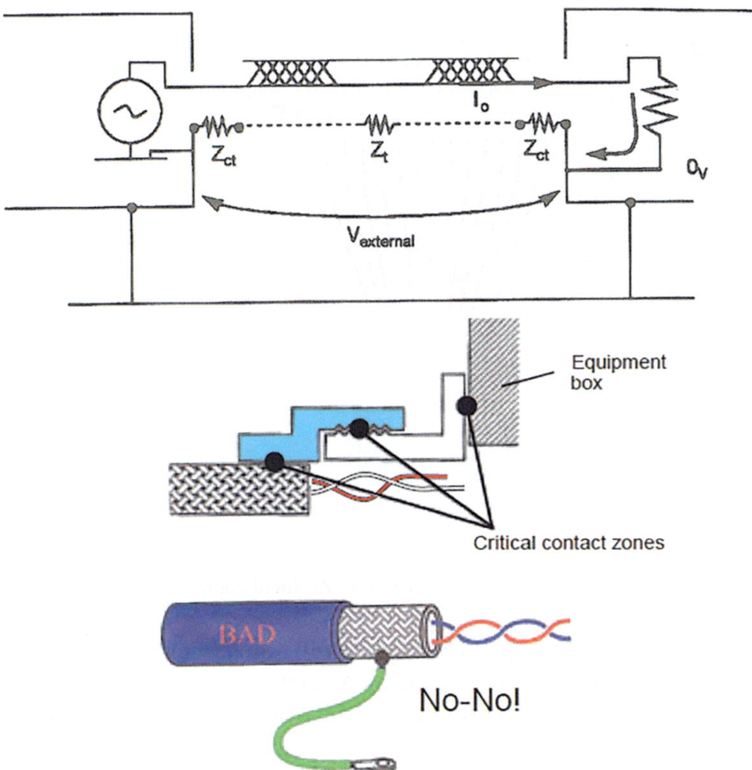

Fig. 5.14 Top: contribution of shield connections impedance Z_{ct} to the global effectiveness. Bottom: practical aspects of a good and bad shield connection, affecting total shield performance

Table 5.1 Typical transfer impedances of *one* shield end connection

	DC to 10 MHz	100 MHz	1000 MHz
Zct, BNC connector	1–2.5 mΩ	5 mΩ	30–100 mΩ
Zct, N conn. (threaded)	<0.1 mΩ	0.03–1 mΩ	0.1–5 mΩ
Zct, SMA conn. (threaded)	<0.05 mΩ	1.5 mΩ	3 mΩ
Ordinary multicontact connector (metal case, sliding, non-threaded)	10–50 mΩ	70 mΩ	300 mΩ
One Pigtail, 5 cm: Zct ≈3 mΩ + j 0.3 Ω × F(MHz)			

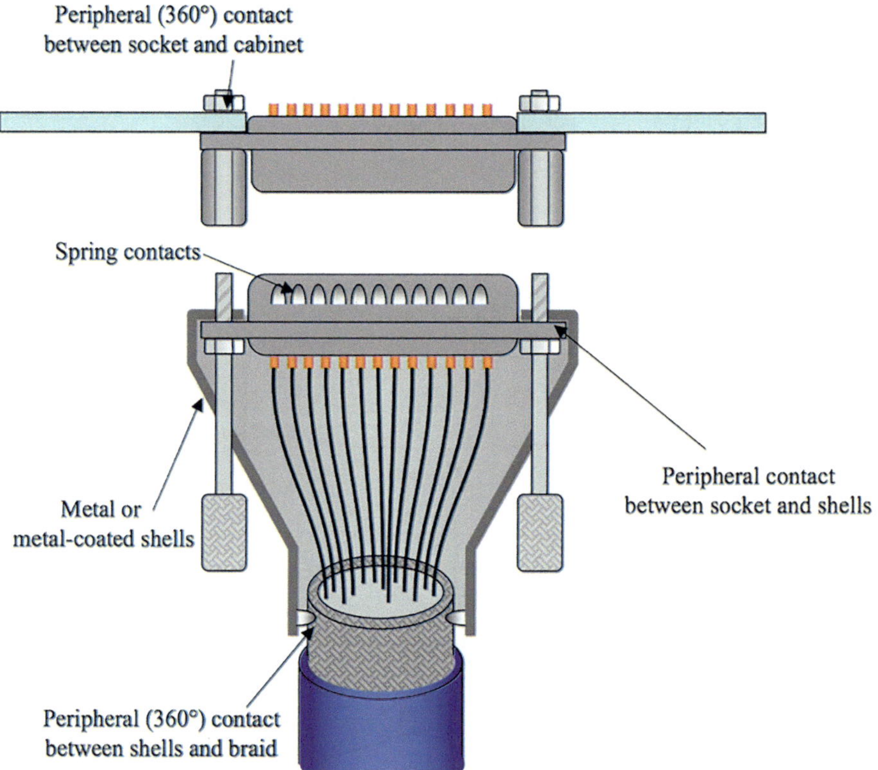

Fig. 5.15 Commercial shielded connectors with low-Zt shield connection

We saw that grounding a shield by pigtails will ruin a shield efficiency. The same is true for internal pigtails gathering the individual shields of STPs. They can pick-up PCB radiation and drive the resulting current over the pairs shields, causing them to radiate. Fixing this simple detail can reduce by 20 dB the emission level. It is vital that a cable shield be terminated by a low-impedance connection (lower than Zt of the cable itself). Most connector styles are available in shielded versions with a 360° contact on the braid (Fig. 5.15).

5.7   Discussion Regarding Shielded vs. Unshielded Twisted Pairs

There is frequent controversy regarding the possibility of using unshielded twisted pairs (UTP) for high speed data links inside buildings, instead of more expensive STP. Measurements and radiation models have shown that in the 30–200 MHz range, a 20–25 dB reduction factor was needed for an ordinary wire pair to satisfy FCC or EN class B. If, instead of the typical 10–30% unbalance of an ordinary single-ended link, a differential link with a high-grade UTP is used, a total 2% unbalance can be achieved, meeting the 20 dB reduction goal. This goal could also be achieved with the help of balancing transformers.

Furthermore, we have seen that the symmetry with a STP cable is slightly inferior to that of the same cable in UTP version. In addition, symmetry impairment is often aggravated by mediocre practices for shield continuity in building wiring. Therefore, many articles are suggesting substantial savings by not using STP. But radiated emission is not the only EMC constraint. In industrial sites, hospitals or high-rise commercial buildings, immunity to RF fields of 10 V/m and to Electrical Fast Transients require added CM protection, that even high-grade UTP cannot provide. Thus, on the basis of immunity, STP is often mandatory in harsh environments.

Quiz
There is ONE good answer or ONE that is better than the others.

1. Coaxial Cables

    (a) A coaxial cable shield should be connected to the signal ground and to the equipment chassis at both ends
    (b) a coaxial cable must be connected to the signal ground only
    (c) A coaxial cable shield should be connected to the signal ground at both ends, and in addition to a good earth rod at the source side
    (d) A coaxial cable shield should be connected to the signal ground at both ends, and to the chassis ground at one end only, to avoid ground loops

2. Shielded pairs (STP) and multiconductors shielded cables

    (a) STP are better than UTP (unshielded) because the shield provide a less resistive path for returning the signal currents
    (b) STP are better than UTP (unshielded) because the shield will collect the EMI currents that otherwise would circulate on the internal wires
    (c) STP are better than UTP (unshielded) because the metallic shield makes a short-circuit for the ground loop voltages
    (d) STP are better because they clearly separate the two functions: the wire pairs carry the intentional current while the shield neutralizes the EMI currents

3. *Shield Transfer Impedance Zt*

   (a) A good cable shield must have a high value Zt
   (b) A good cable shield must have a low value Zt
   (c) Zt can be calculated from the coaxial cable or STP characteristic impedance
   (d) Shield Transfer Impedance provide a measure in dB of the cable shielding effectiveness

4. Calculate shield induced noise by Zt: a 5 m length of RG58 coaxial cable is exposed to an RF field that induces 30 V at 10 MHz in the cable-to-ground loop. The loop impedance at 10 MHz is $\approx 150\ \Omega$. What is the noise voltage appearing inside on the center conductor?

   (a) 4 V
   (b) 80 mV
   (c) 0.2 mV
   (d) 8 mV

Chapter 6
Shielding of Boxes and Enclosures

Preamble

Former chapters reviewed the principal conduction and radiation coupling mechanisms, as they affect equipment/system susceptibility, and the last one was addressing shielded cables. This chapter is focusing on the shielding of equipment boxes, from the smaller hand-held devices up to large cabinets or even entire rooms.

Why addressing separately cables shielding and box shielding? Against an ElectroMagnetic field a shield is a shield, no matter if it is a tube or a cube closely coupled with a tubular envelope, such as it is the mutual inductance that does the canceling effect. With a shielded enclosure there is no such close coupling: it is the portion of the field which goes through the barrier that gives a measure of the shield effectiveness.

6.1 Do Not Let Shielding Happen by Chance: Design for It

Once EMI reduction techniques described earlier have been applied (PCB layout, equipotential grounding, ground loop isolation, loop area, etc.), a conductive box may be the ultimate barrier against radiated susceptibility or emissions. However, too often, EMC performance is not regarded as a key element by people designing equipment housing. One or a combination of the following approaches are used instead that could be captioned "recipes for EMC failure":

- Make the enclosure similar to earlier versions that were deemed to be EMI-free. Then, confirm expectations by prequalification tests on a functional prototype.
- Starting from the ground up, make a box according to mechanical, esthetic, cost, and accessibility criteria, then test it as above.
- Do as above, but perform only the mandatory radiated emission tests. Do not test for susceptibility unless a purchasing specification calls for it.

© The Author(s), under exclusive license to Springer Nature
Switzerland AG 2025
M. Mardiguian, *ElectroMagnetic Compatibility*,
https://doi.org/10.1007/978-3-032-02688-0_6

Such regrettable hit-or-miss process means that *it is the final test that governs the outcome of a design*, resulting in one or more of the following:

- Time and money are wasted during the iterations.
- Components or techniques that are not optimized become integral parts of the product.
- EMC overdesign, with its accumulation of cost, weight, and maintainability issues.
- EMC underdesign because system tests may not cover all possible EMI situations.

Instead, the designer who chooses an analytical approach should consider the following:

(a) How much attenuation (if any) should the enclosure provide?
(b) How to design an enclosure meeting the attenuation goal before a prototype exists?
(c) If item (a) is not known, as is usually the case, how can it be quantified?

Thus, a deterministic EMC approach ("Make-it-Right-the First Time") to the design of an enclosure is needed. Emission and susceptibility being related, we will address the combination of both.

6.2   Quantifying the Need for Box Attenuation

Universally agreed for the performance of a shield, shielding effectiveness (SE) is the ratio of the incoming field to the residual outgoing field (the part that gets through the barrier).

$$\text{for E} - \text{fields} : \text{SE}\left(\text{dB}\right) = 20\log E_{in} / E_{out} \quad \text{for H} - \text{fields} : \text{SE}\left(\text{dB}\right) = 20\log H_{in} / H_{out}$$

Notice that a more rigorous definition would be the ratio of the **field that existed** in a specific point in space **without the shield**, to the field that remains once the shield is in place. Compliance to this definition is seldom feasible in a practical test (Fig. 6.1).

Note

Quick reminder on the deciBel:

SE (dB)	Corresponding E or (H) attenuation ratio
6	2
10	3.16
20	10
40	100
60	1000

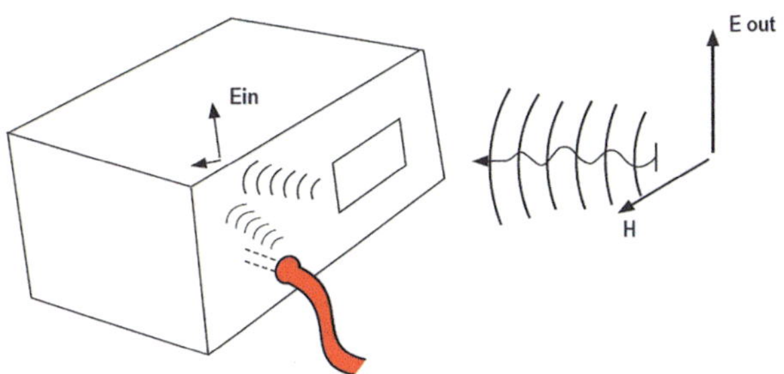

Fig. 6.1 RF penetration can occur through the box walls, apertures (cooling vents, access holes, poorly joined assembly, etc.) and cable penetrations

Using Fig. 6.2 routine, we first ask whether the required shielding is already known across a defined frequency range. While it is generally not known, cases occur where procurement specs or test data from a similar equipment dictate the necessary degree of SE. In such case, the SE routine is bypassed, except for adding a safety margin.

Thus, assuming the needed SE is generally unknown, the flow diagram covers three cases:

(a) ***Shielding for immunity hardening***

- Determine the ambient threats (LF magnetic field, electric field, etc.), frequency and amplitude. This is based on the product's intended application/location and found in the immunity specs. For a new application, if no specification exists, a site survey is needed.
- Compute, or evaluate via a prototype, the interference situation via the coupling of fields to internal cables and PCBs. This includes in-band and out-of-band response of victim circuits.
- The desired SE is the ratio in dB between the imposed threat and the "barebones" susceptibility of the unshielded equipment.

(b) ***Shielding for emission control***

- Compute, or measure on development prototypes, the radiated emission levels for each major subassembly to be housed in the box, excluding I/O cables whose radiation will be addressed and resolved separately from box shielding. For each frequency interval of at least one decade (half-decade intervals are preferred), record the highest calculated/measured field level up to $\approx 10 \times F_2$. F_2 being the highest significant frequency of the functional voltages or current spectra, for inst. $0.35/t_r$ for pulsed signals.
- If several amplitudes are in the same range, compute their combined effects. Once a radiated field envelope is drawn across the spectrum for the unshielded electronics, it is compared to the applicable civilian/military Rad. Emission specification to define the SE.

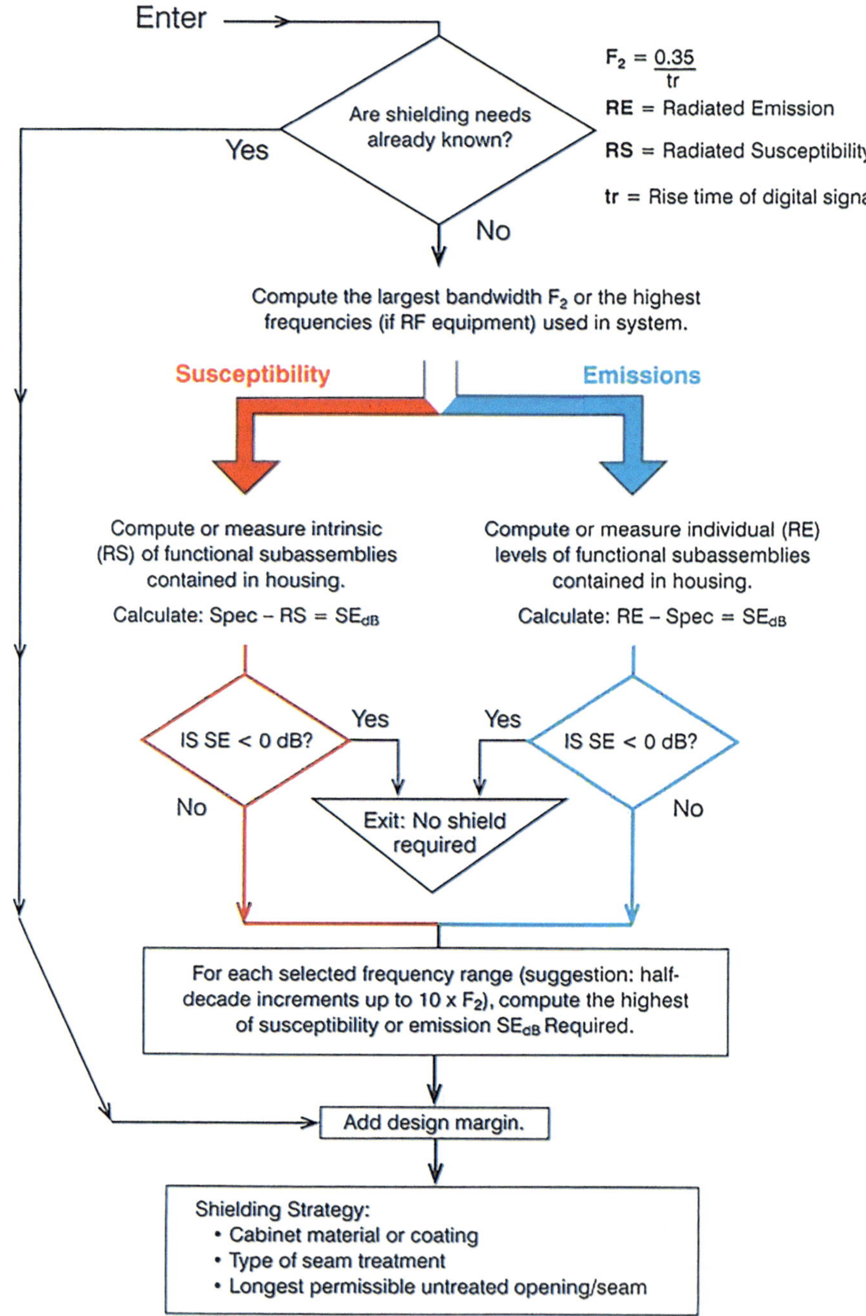

Fig. 6.2 Flow diagram for shield design. The right-hand branch emphasizes shielding against radiated emission

(c) ***Optimized shielding for dual Susceptibility and Emission control.***

Once the SE (a) and (b) have been determined, compare the susceptibility SE(a) and emission SE(b) in each frequency interval, to retain the toughest of the two requirements. "Toughest" is not necessarily the highest dB figure. For instance, 20 dB SE against a near-field magnetic source may be harder to achieve than 60 dB against an E-field or plane wave at the same frequency.

Having established the final SE requirements, what remains is to select or verify:

- the cabinet material
- the way apertures and seams will be treated
- the surface treatment/finish if corrosion and longevity requirements exist.

For emission control, the constant increase in clock frequencies >1 GHz or higher spectrum obliges to consider possible leakages from any slot whose dimension exceeds a few centimeters. Decades ago, empirical methods often led to a "hammer" approach where equipment housings resembled a vault. Although effective, it is adding manufacturing and hardware costs, complicate maintenance or accessibility. Sometimes, esthetic and weight considerations prohibit the use of certain shielding materials.

Today, designers are looking for shielding techniques that are economical and resist to intensive use. For consumer devices, immunity SE requirements are usually less demanding than those for emission, because radio-protection limits are more stringent for residential areas. Conversely, for military, aerospace or automobile equipment, immunity requirements may be the leading constraint. Even with no or poor EMC design of the PCB and internal packaging (meaning the shield will have to make up for internal deficiencies), SE in the 10–40 dB range for civilian applications, and in the 30–60 dB range for MIL-461, are generally adequate at the worst offending frequencies.

6.3   Basic Mechanisms of Shielding

A deep coverage of shielding theory is far beyond our scope, but major guidelines are provided on how and why shields work and eventually why they do not. Readers willing to know more about shielding can find clear and concise summaries in [6, 8, 10]. More complete theory is found in [7, 11].

Shielding can be regarded as the result from a *brutal encounter between two actors*: an electro-magnetic field and a conductive barrier (most of times metallic). Understanding how a shield works implies that we know these two actors.

6.3.1 The Nature of the Incident Field (Intentional, or Incidental)

Depending on how it has been created and how far the source from the barrier is, three situations may exist (Fig. 6.3):

- **Far-field conditions** (also known as *"plane waves"*): the source is far from the barrier, at a distance $d \geq \lambda/2\pi$ or $d(m) \geq 48/F(MHz)$. In this case, the field parameters are stabilized and well characterized:

  - Wave impedance $E(V/m)/H(A/m) = 377\ \Omega$, forever.
  - Field fall-off rate is $1/d$, for both E and H terms.
  - The magnitude of the field is always expressed by its "E" term, in V/m. If the true value of the associated H-field is needed, we should compute $H = E/377$. For instance, in far-field, a 1 V/m E-field is associated to a H-field of $2.7 \cdot 10^{-3}$ A/m.

- **Near-field conditions:** the source is close to the barrier, at a distance $d \leq \lambda/2\pi$, translated as $d(m) \leq 48/F(MHz)$. Per Fig. 6.3, two situations may exist:

  - *Near-field Case (a) field is predominently Electrical* (V/m), if it has been created by an open wire (dipole or rod), or any high impedance circuit. It has a low value of associated H-field (A/m), and is regarded as a *high impedance* one, because the ratio E/H has a large value in Ohms. The E-field falls off like $1/d^3$.
  - *Near-field Case (b) is predominently Magnetic(A/m)* if created by a closed loop (magnetic antenna), or any low impedance circuit. The Amplitude of the

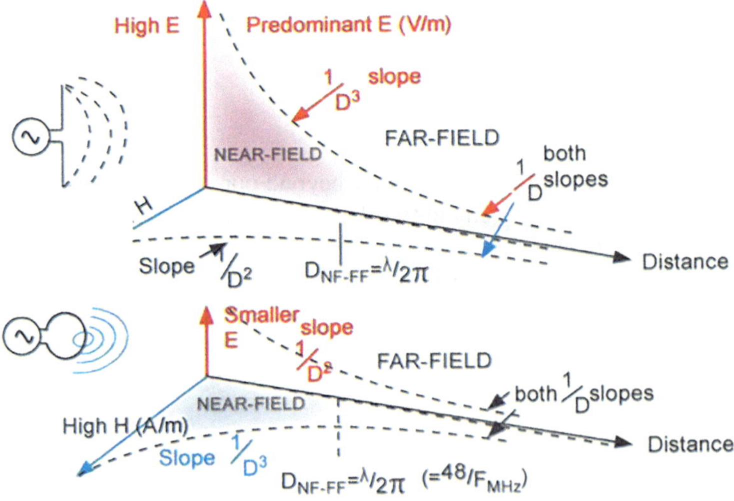

Fig. 6.3 Near-field for electric or magnetic excitations, and far-field, with transition distance $D_{NF\text{-}FF}$

associated E-field (V/m) is low, and it is regarded as a *low-impedance field*, since the ratio E/H, is a low value in Ohms. Such near-field condition is hard to control by shielding. The H-field falls off like $1/d^3$.

6.3.2   The Characteristics of the Barrier

Shielding properties are dictated by the material resistivity (or its inverse: conductivity) and magnetic permeability. These two constants are determining the surface impedance and penetration depth of the barrier at a given frequency. If shields were perfect the emerging fields E_{out}, H_{out}, hence output power P_{out} would be null. But a shield is more an attenuator performing by two principles: reflection and absorption (Fig. 6.4).

(a) ***Reflection loss:*** *For far-field conditions*, reflection loss is given by:

$$Rd(dB) = 20\log(K+1)^2 / 4K \quad \text{where } K = 120\pi / Z_b \qquad (6.1)$$

For K > 3, Eq. (6.1) simplifies as: $R(dB) \approx 20 \, Log(100/Z_b)$ where Z_b = barrier impedance at the interface
Near-field conditions, where the shield is closer than $\lambda/2\pi$ to the source, are the most critical.

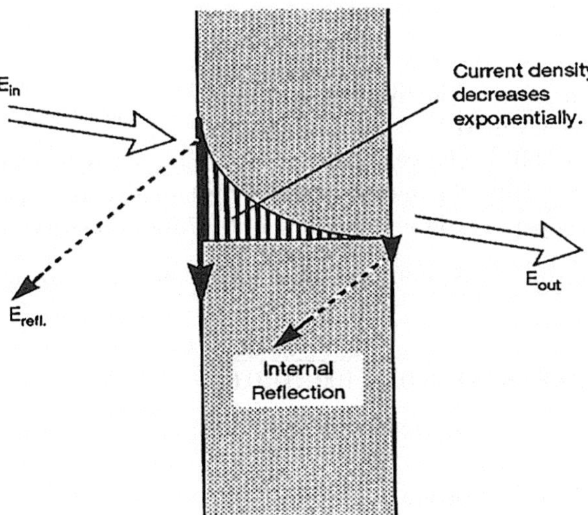

Fig. 6.4 Basic shielding mechanisms: *reflection increases* with surface conductivity and wave impedance. *Absorption loss increases* with thickness, conductivity, permeability and frequency

Against predominent **E-field**, the high wave impedance provides easily good reflection factor, thanks to the field-to-shield impedance mismatch. Think of a field impedance >1 kΩ meeting a barrier impedance <1 Ω. As frequency increases (decreasing λ), the high impedance of the field falls off until far-field conditions are reached, when distance $D \geq \lambda/2\pi$.

Against nearby **H-fields**, wave impedance is low, and it is more difficult to get good reflection, so the results are just the reverse. Near H-field reflection loss is given by:

$$R\left(dB\right) = 20\log 2D.F / Z_b \tag{6.2}$$

Note   R(dB) is an *attenuation*, never a gain and cannot be negative: thus, when $2D.F/Z_b$ becomes <1, R must be clamped to 0 dB.

How does one know if at distance $\ll\lambda$, the field is more electric or magnetic? Looking at the radiating source give an idea of the predominant mode: sources switching large currents such as power supplies, solenoid drivers or large ICs handling more than 100 mA/V create predominant H-fields. Conversely, voltage-driven high-impedance or open-ended lines create electric fields.

(b) *Absorption Loss*

To evaluate *absorption,* or penetration losses, we need to know how many skin depths (δ) the metal barrier represents at the frequency of concern; field intensity will decrease by 8.7 dB (or lose 63% of its amplitude) each time it goes through one skin depth. Entering all electrical constants, we come to a simple expression for absorption loss:

$$A\left(dB\right) = 131t\sqrt{F\mu_r \sigma_r} \tag{6.3}$$

where

t = thickness of conductive barrier in mm
F = frequency in MHz
μ_r = permeability relative to copper = 1 for non-magnetic materials (Fig. 6.5)
σ_r = conductivity (the inverse of resistivity) relative to copper
= 1 for copper, $\approx$ 0.6 for aluminum, or 0.17 for ordinary construction steel

For example, the absorption loss of a 0.03 mm (1.2 mil) aluminum foil at 100 MHz is:

$$A\left(dB\right) = 131\times0.03\sqrt{\left(100\times1\times0.6\right)} = 30.4\,dB$$
$$= \text{a reduction factor of } \left(10\right)^{30.4/20} = 33\,\text{times.}$$

Looking at Eq. (6.3) leads to a few remarks:

1. *For non-magnetic materials* ($\mu_r = 1$), penetration losses increase with conductivity σ_r. Since no metal has better conductivity (except for silver, where $\sigma_r = 1.05$),

Material	Relative Conductivity (σ_r)	Relative Permeability (μ_r)	Skin Depth (δ)			
			50 Hz	10 kHz	1 MHz	100 MHz
Copper	1	1	9.3 mm	0.66 mm	66 µm	6.6 µm
Aluminium	0.6	1	12 mm	0.85 mm	85 µm	8.5 µm
Steel (ordinary cold-rolled)	0.16	200	1.5 mm	0.14 mm	100 µm	16 µm
Hi-perm. alloy(typ)	0.03	10 000	0.54 mm	54µm	*NA	*NA

(*) μr of Hi permeability alloys is generally collapsing above a few MHz.

Absorption losses (dB)

	Copper			Aluminium			Zinc			Steel			Copper Paint
Thickness (mm)	0.01	0.1	1	0.01	0.1	1	0.01	0.1	1	0.01	0.1	1	0.05
30 MHz	7	70	700	5.2	52	520	4	40	400	3	28	200	7
100 MHz	13	130	>1000	9.5	95	950	7	72	720	5	50	500	13
300 MHz	22	220	>1000	17	170	>1000	12	125	>1000	9	88	880	22

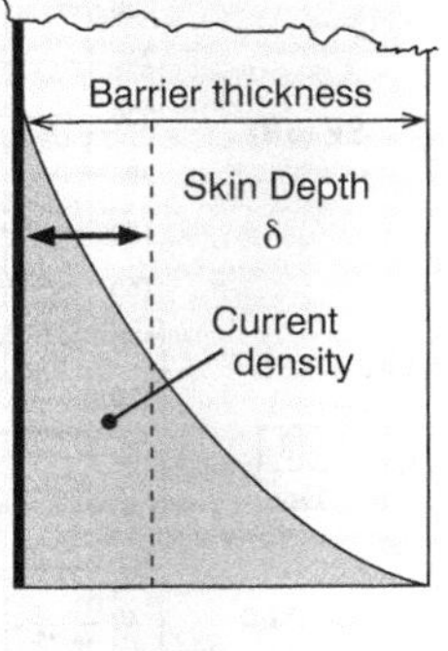

Fig. 6.5 Skin depths and absorption losses for various materials (reminder: absorption does not depend on the near-electric, near-magnetic or plane wave type of field)

any non-magnetic metal will show less absorption than copper. Zinc, for instance, with $\sigma_r = 0.3$, for a same 0.03 mm thickness has an absorption loss at 100 MHz of:

$$A(dB) = 131 \times 0.03 \sqrt{(100 \times 1 \times 0.3)} = 20\,dB$$

2. *For magnetic materials* ($\mu_r > 1$), penetration loss increase with μ_r, but, their conductivity is less than copper. Yet, with μ_r for steel or iron in the range of 300–1000, while σ_r is about 0.17, a clear advantage exists for magnetic materials. Above a few hundred kilohertz (ferrites excepted) μ_r generally starts collapsing to 1, while σ_r is still mediocre. Figures 6.5 and 6.6 show shielding properties of some common metals.

Note Although Reflection and Absorption are presented as two independent factors, they are interacting. Reflection on the air-to-metal interface combines with the internal absorption, followed by a reflection on the second metal-to-air interface, that in turn is altered by the multiple internal reflections.

Reflection Eq. (6.3) is taking into account these in-between mechanisms. However, *for a thin barrier* with thickness (t) is < skin depth (δ), no absorption exists: shielding is entirely due to the barrier reflection, without the multiple internal reflections described above. *In this specific case* [1, 9] shielding by reflection loss is given by:

$$\text{Thin wall, } R_{(dB)} = 20 \log(120\pi / 2Z_b) \approx 46 - 20 \log R_b \qquad (6.4)$$

where R_b = surface resistance (dc) of the thin film, in Ω/sq.

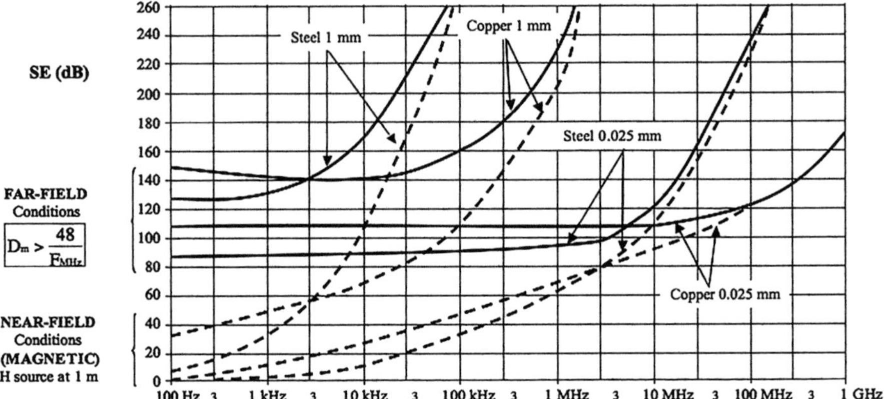

Fig. 6.6 Total shielding effectiveness (absorpt. + reflection) of common metals. **Solid lines** (top): far-field conditions. **Dotted lines** (bottom): for H-field sources at 1 m distance. Curves for 1 mm copper can be used for 1.25 mm (0.05″) aluminum. Curves for 0.025 mm copper (1 mil) can be used for 0.03 mm aluminum. (From [11])

Table 6.1 Average performance of conductive treatments on plastics, compiled from several sources (Acheson, MAP, Parker/Chomerics). Graphite works only against high impedance, E-field shielding

	Surface resistance, Ω/sq	SE(dB), far-field
Copper paint (50 μm thick)	0.2	60
Nickel paint (50 μm thick)	0.3–0.08	56–48
Graphite paint (50 μm thick)	10–300	26 to ≈0
Silver coating (13-25 μm thick)	0.06	70
Hot zinc spray	0.05	72
Copper, electroplating	0.1	66
Copper, electroless plating	0.1	66

6.3.3   Shielding Effectiveness of Conductive Plastics

Plastic housings provide basically no shielding. Thus, unless the inside circuitry has been hardened, the plastic must be made conductive in order to provide some shielding. Metallizing processes exist, summarized in Table 6.1, with cost ranging from 10 to 200\$/m². Since, as aforesaid, thin coatings exhibit poor or no absorption, they only work by reflection. Based on this, the table shows the SE of thin coatings. If a SE in the range of 40–50 dB is desired, a process with a surface resistivity 1 Ω/sq. or less must be selected.

6.4   Shield Degradation Caused by Apertures

All the SE values given above assume a plain, homogeneous barrier. Real-life housings are never continuous metal cubicles: slots, seams, and other apertures will inevitably leak. As for a chain, a shield is only as good as its weakest link; therefore it is paramount to know the shield's weak points in order to match realistic objectives.

- At low frequencies, what counts is the type of metal (conductivity, permeability) and its thickness.
- At high frequencies, any metal would provide SE > hundreds of decibels, but such figures are never seen because seams and discontinuities are spoiling the metal barrier (Fig. 6.7).

6.4.1   Attenuation of One Single Aperture

A slot in a shield can compare to a slot antenna which, except for a 90° rotation, behaves like a dipole. When slot length reaches $\lambda/2$, no matter how small the height (h), *this unintentional antenna acts as a perfectly tuned dipole*, that is, it re-radiates on the "exit" side all the energy that excites the slot on the incoming side. It may even exhibit a $\approx$3 dB gain. Below this resonance, the slot leaks less and less as frequency decreases. The equivalent circuit for a slot is an inductance (Fig. 6.8), until it resonates with the edge-to-edge capacitance, when $l = \lambda/2$. Maximum leakage occurs for the cross polarization of the slot (E-field vertical illuminating an horizontal slot).

A simplified expression gives aperture attenuation below $\lambda//2$ resonance [9]. It is the worstcase far-field attenuation, for the worst possible polarization (in general, actual attenuation will be better).

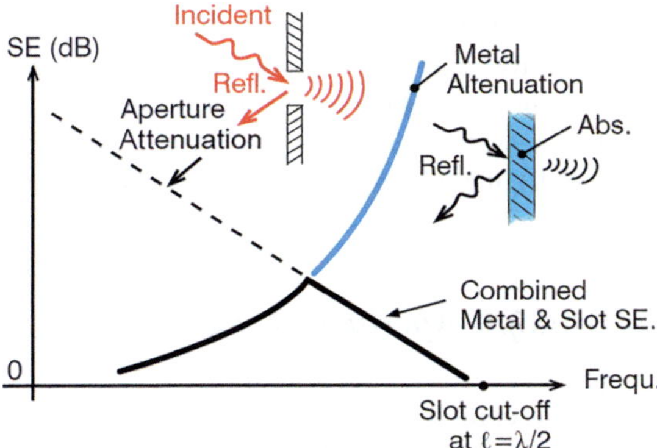

Fig. 6.7 Field attenuation by a plain metal barrier (increasing with F) compared with the attenuation through an aperture (decreasing with F)

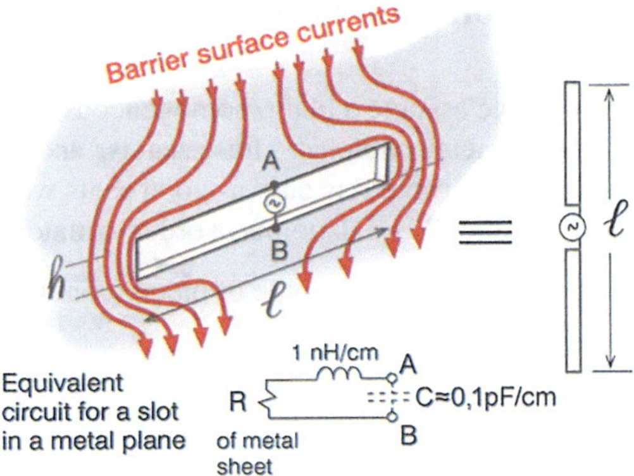

Fig. 6.8 Effect of a discontinuity in a shield. A slot with length l behaves as a dipole of same length. At low frequency, slot impedance Z_{AB} is shorted by the low metal resistance: reflection is significant

$$A\,(dB) \approx 100 - 20\log l - 20\log F\,(MHz) + 20\log\big[1 + 2.3\log(l/h)\big] + 30\,d\,/\,l$$
$$[\text{-----------------------------------}]$$
$$= 0\,dB \text{ for } l \geq \lambda\,/\,2 \qquad\qquad (6.5)$$

with l = length (largest dimension) of aperture in mm.

h, d = height and depth of the aperture (no unit given since it is the ratio l/h or d/l).

The first three terms in Eq. (6.5) represent the reflection loss of a square aperture, due to the mismatch of the incident wave impedance (377 Ω for far-field conditions) with the slot impedance.

The fourth term is the "fatness factor" of the slot, accounting for the effect of h. Notice that h plays only a secondary role by the logarithm of l/h. A slot 100 times thinner will not radiate 100 times less than the equivalent square aperture, but only 5 times less (see Table 6.2).

The last term, 30 d/l, directly given in dB, is the guided wave attenuation, as it would happen in a real waveguide below its operating frequency. It has only some effect if depth "d" is a significant fraction of l. For ordinary sheet metal where "d" is simply the metal thickness, this term is negligible. For small holes, or artificially lengthened holes (Fig. 6.9), the added attenuation is substantial.

6.4.2   Effect of Multiple Apertures Leakages

How can we estimate the combined effect of several apertures, whether they are similar or not? Following are some guidelines for the most frequent cases.

(a) **Several apertures, scattered and *not identical*:**

Table 6.2 Attenuation (far-field) for a few square apertures. For rectangular slot, use correction (fatness factor) on right-hand table. Attenuation fall to 0 dB when *l* reaches half-wavelength, *regardless the height "h"*

ℓ

Aperture size	30cm	10cm	3cm	1cm
Freq. Atten, (dB)				
100kHz	70	80	90	100
1Mhz	50	60	70	80
3MHz	40	50	60	70
10MHz	30	40	50	60
30MHz	20	30	40	50
100MHz	10	20	30	40
300MHz	≈0	10	20	30
1GHz	0	≈0	10	20
3GHz	0	0	≈0	10

Added correction for $\ell > h$

$\ell = h$ Δ = 0dB
$\ell = 10h$ Δ = +10dB
$\ell = 100h$ Δ = +15dB
$\ell = 1000h$ Δ = +18dB
Correction Δ disappear when $\ell = \lambda/2$

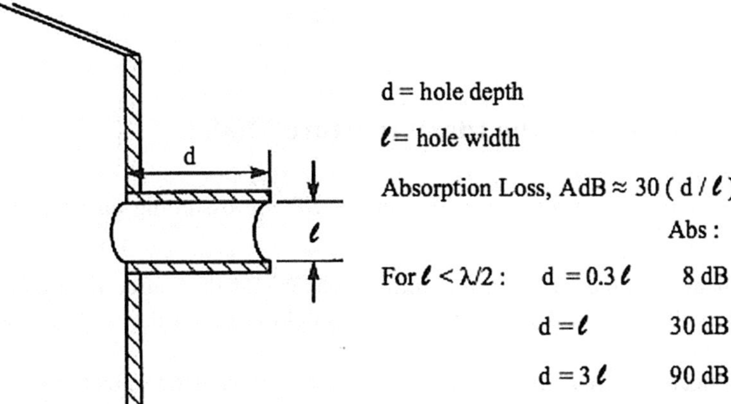

d = hole depth

ℓ = hole width

Absorption Loss, AdB $\approx$ 30 (d / ℓ)

Abs :

For $\ell < \lambda/2$:   d = 0.3ℓ 8 dB
             d = ℓ 30 dB
             d = 3ℓ 90 dB

Fig. 6.9 Additional attenuation offered by lengthened holes (waveguide below resonance)

Compute A(dB) for each aperture. The global SE will be approximately that of the poorest attenuation(greatest leakage). A more accurate prediction can be made by *adding the leakages*, NOT the dBs.

Example: *assume two apertures, where calculations using* Eq. (6.5) *has given:*
Aperture #1: 26 dB (a relative leakage of 0.05, meaning 5% of field gets through),
Aperture #2: 14 dB (a relative leakage of 0.2, meaning 20% of field gets through),
The total leak is: 0.2 + 0.05 = 0.25, that is a total attenuation = −(20 log 0.25) = 12 dB. As predictable, the total attenuation is slightly less than Ap#2, the worst one.

(b) **N apertures, identical but scattered (*not adjacent: t ≫ length L*):**

Compute A(dB) for one hole, and *substract* 20 Log N. This gross worstcase assumes that all openings are re-radiating in phase, which is not entirely true. If there are many apertures, such as the global result is approaching 0 dB, total SE must be clamped to 0 dB: slots cannot cause negative loss and amplify the field!

(c) **N apertures identical, not scattered and *adjacent,* with thickness of ribs t ≪ L or h,**

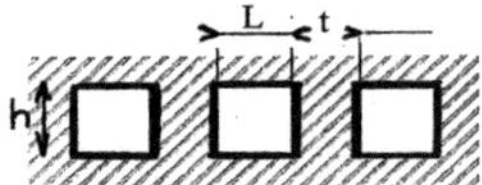

Compute A(dB) for *only one* aperture. DO NOT substract for multiple holes. This is because when identical holes are separated by thin ribs, mutual cancelation occurs by the edge currents (see Fig. 6.10, bottom).

6.5   Alterations of the Ideal Aperture Model

Former calculations for metal and aperture SE are assuming an academic situation where:

- the metal wall has quasi infinite dimensions versus the source-to-shield distance such as the current density in the plain shield (without any leakage) would be uniform,
- the reflected wave does not hit any opposite wall, causing reflections.

  Reality is different:

(a) Cabinets have finite dimensions, causing current concentrations at the edges.
(b) the box may behave as a cavity excited by internal sources, if one or several of the circuit frequencies meet the natural box resonances.
(c) For radiated emissions, sources can be rather close from the box walls openings, such as a good part of the concerned frequency range is in a near-field condition.

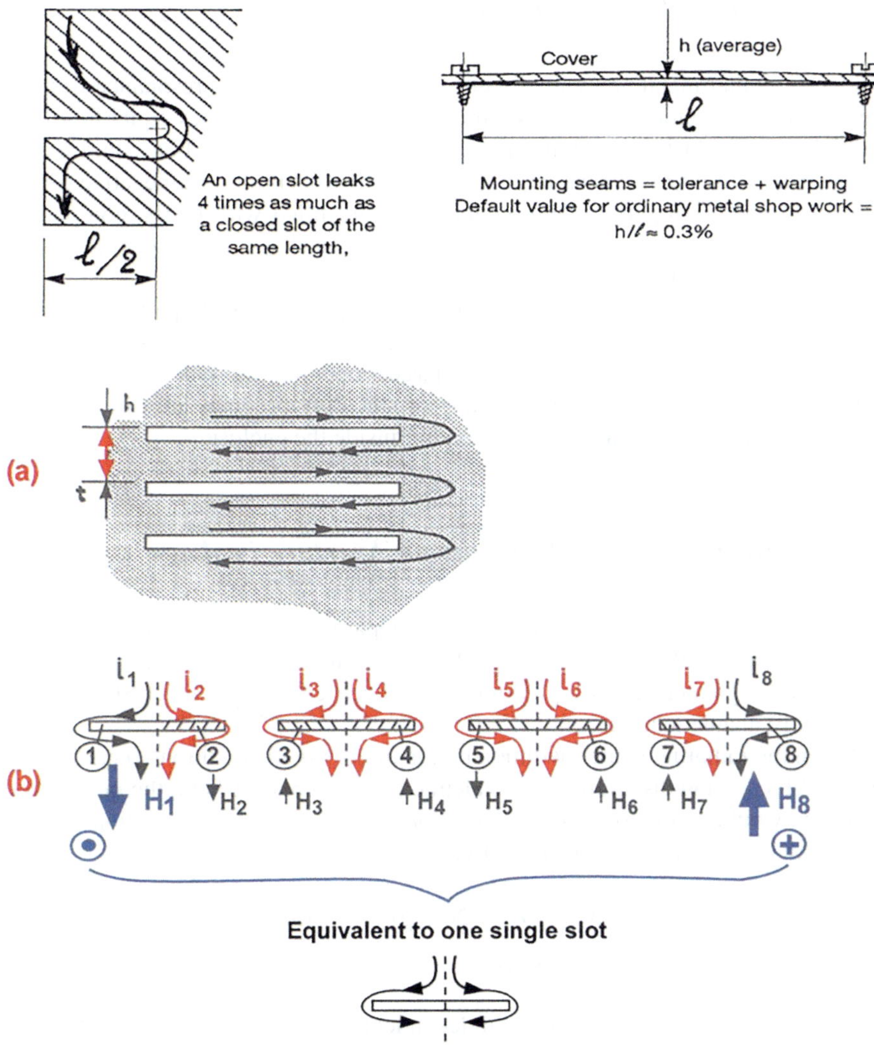

Fig. 6.10 Some typical box leakages. With (**a**), the metal "t" between adjacent slots is larger than the slot height "h", so the slots radiate like independent antennas. With (**b**) the thin rib between contiguous slots is causing all half-slots pairs like (2) + (3), (4) + (5), etc. to act as *opposite magnetic moments*: seen at distance, their radiations are mutually canceling. Only the radiations of the first and Nth half-loops are combining *like if it was a single slot*

6.5.1   *Effect of Box Natural Resonances*

For a rectangular metal box with dimensions l, w, h, the natural resonance frequencies of the waveguide occur every time the path length becomes equal to an odd multiple of $\lambda/2$. At these specific frequencies, an empty metal box could exhibit resonances with a Q factor as large as 10 (20 dB), resulting in a negative SE (an apparent "gain"). Hopefully, equipment boxes are filled with PCBs, components and cables behaving as scattered lossy elements, such as the measured Q stays within 0–10 dB, with typical values of 6 dB.

This doubling of the inside field results in an apparent 6 dB drop of the expected SE at every self-resonance frequency. If box self-resonant frequencies are less than twice below the cut-off frequency of its largest opening, the box SE will drop to 0 dB faster than expected. Some vendors of shielding products are offering lossy ferrite mix that can work as anechoic coating inside the enclosure.

6.5.2   *Effect of Source Proximity on Aperture Leakage*

As mentioned, aperture SE in the near-field departs significantly from its far-field value of Eq. (6.5). Wave impedance will differ from 377 Ω, affecting the reflection term, hence a greater SE for a predominantly electric field, but lower SE for a predominantly magnetic field.

When the radiating source within a box is in near-field conditions (i.e., distance $D_{(m)} < 48/F_{(MHz)}$) and discarding the academic case of a pure E-field, the most severe condition would be that of a near-field H source. At worst, the attenuation of a slot against an ideal H-field loop is given by:

$$
\begin{aligned}
\mathrm{SE}\left(\text{near H - field}\right) &= 20\log D\,/\,l + 20\ \log\left[1 + 2.3\log\left(l\,/\,\mathrm{h}\right)\right] + 30\,\mathrm{d}\,/\,l \\
&= 0 \text{ dB for } l \geq \lambda\,/\,2
\end{aligned}
\tag{6.6}
$$

Notice that this last expression is independent of frequency, as long as the near-field criterion and quasi perfect H-field exist. It can be regarded as the worst low boundary of aperture SE against H-field sources.

6.6   Reducing Apertures Leakage for a Given Shielding Objective

A conductive housing is naturally an efficient barrier. The designer should aim at not spoiling this barrier with excessive leakages. Leakages (i.e., poor SE) are caused by:

- Assembly seams at mating panels, covers, etc. (a frequent cause of spoiling SE).
- Cooling apertures.
- Viewing apertures.
- Component holes: fuses, switches, keyboards, shafts.
- Cable or miscellaneous conduits penetrations.

6.6.1   Mating Panels and Cover Seams

The general, simple rule is: all metal parts of an housing should be bonded together: a floated item is a candidate for re-radiation.

For cover seams, slots, and so forth, how frequently they should be bonded depends on the design SE objective. Table 6.2 provides that a 10 cm leakage is worth $\approx$20 dB of shielding at 100 MHz. If the goal is closer to 30 or 40 dB, seams or slots should be broken down to 3 or 1 cm. For permanent or semi-permanent closures, this means many screws or welding points. For covers, hatches, and such, flexible contacts or gasket are preferred. In any case, a safe practice is to design fold-over shapes to the cover edges. With a large overlapping, a waveguide "labyrinth" is formed, adding some penetration loss. This could complement the use of gaskets, or even avoid them.

The following is a sequential list of these solutions in ascending order of efficiency and cost.

- *If only minimal SE is needed, in the 0–20 dB range*, simplest technique is to have frequent bonding points and, for covers, short straps made of flat braid or copper foil as shown in Fig. 6.11. This is bonding only on the hinge side but can be sufficient if no sensitive or noisy cables or devices are located near the opposite side of the hinge. For the unbonded opposite side, use grounded locks/fasteners. The λ/20 criterion means that, for a maximum EMI frequency of 100 MHz, the maximum distance between ground straps is 10 cm for a 20 dB shielding objective. However, for emission shielding, this would imply that the emission source inside be at a distance >10 cm from the leaky seam, which may not be the case (see previous "effects of source proximity").
- If bonding only the hinged side leaves too much of leaky seam, more bonding points are necessary. In this case, soft springs can be scattered along cover edges (Figs. 6.12 and 6.13). For durable performance, the spring contact riveting must be corrosion free. Many types of contacts fingers are available, such as low-

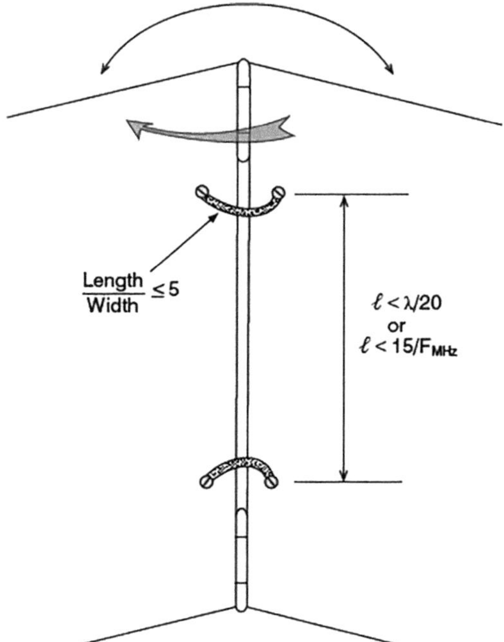

Fig. 6.11 Leakage reduction by scattered seam bonding (for moderate shielding needs)

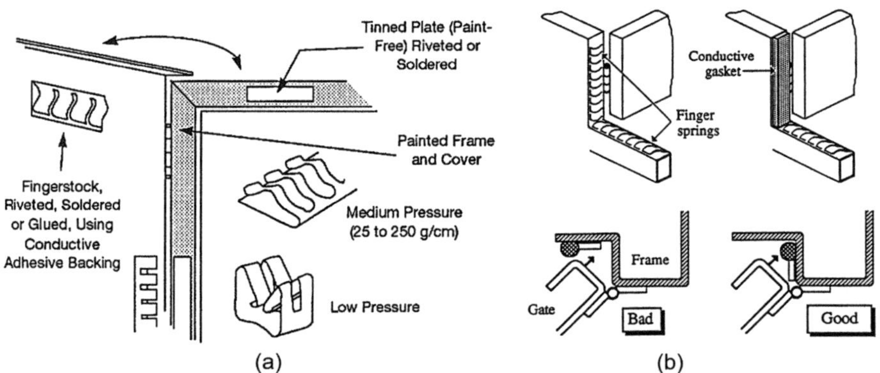

Fig. 6.12 (**a**) Maintaining shield integrity by evenly spaced, flexible bonding points. (**b**) Full shield integrity by continuous contacting

pressure, knife edge and medium-pressure styles. Provided an adequate control of pressure through tight mounting tolerances, they are very dependable. The technique shown in Fig. 6.13 is taking minimal surface preparation. Grounding buttons, mounted by press fit or threaded stud are compliant to gap variations.

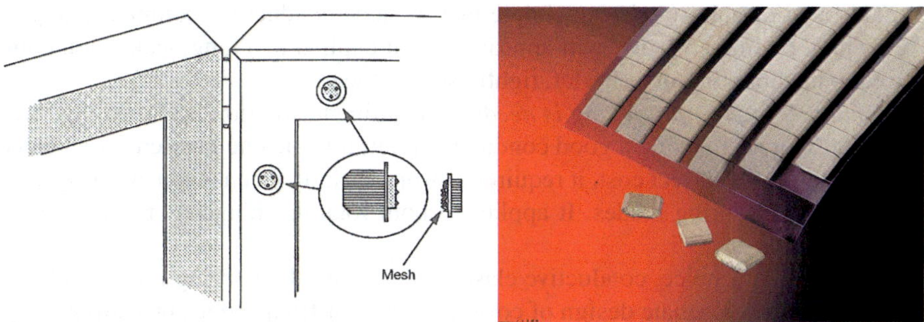

Fig. 6.13 **Left**: press-fit grounding buttons (LAIRD Co.). **Right**: soft grounding pads (Chomerics Parker div)

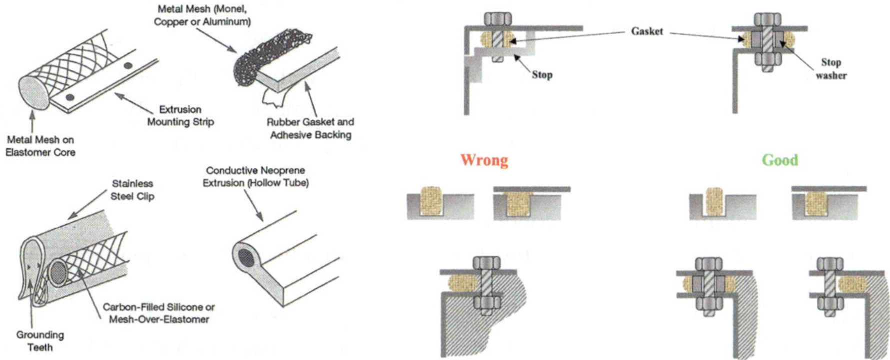

Fig. 6.14 Compressible RF gaskets and mounting styles. Gaskets should be sufficiently pressed, but not over-pressed to the point of losing their elasticity

- *If a higher shielding is required (20–60 dB)*, a continuous conductive bonding of seams is necessary, since an SE of 40 dB at 300 MHz ($\lambda//2 = 50$ cm) would require screws or rivets spacing of 0.5 cm! Continuous conductive joints are available in many forms and stiffness (Fig. 6.14). Mesh-type gaskets provide higher shielding, close to or beyond the upper side of the required SE range. Hollow elastomer gaskets are inexpensive because their elasticity compensates for large joint unevenness and warpage. The counterpart is a lesser contact pressure, hence higher resistivity; it is a solution for the lower side of the SE range.

A Word of Caution Regarding EMC Gaskets Performances

Gasket performances (SEdB) shown by vendors are only valid for a specific test set-up.

This is not a sales pitch, but simple fact. A same conductive, compressible gasket, with the same pressure, will appear better if applied to a long, leaky slot, compared to its mounting in a shorter, tightly secured seam.

Finally, if highest hardening is needed, the ultimate solution is shown Fig. 6.14 with 100% of seam being a good conductive joint; it is the one favored for shielded rooms. Besides its direct cost, it requires a strong locking mechanism ensuring even pressure on all spring blades. It applies to both rotating (hinged) or slide-mating surfaces.

Whatever the choice, conductive elastomer core, mesh or spring fingers, all gaskets requires an adequate design of covers and box or frame edges providing:

- a smooth seating plane or groove for the gasket, with well conductive surface finish,
- tight mechanical tolerances avoiding gasket overpressure at some places (lower tolerance gap) causing gasket crushing, and underpressure at others (higher gap) resulting in insufficient contact.

In any case, a good quality, corrosion-free mating surfaces with good conductivity is mandatory. This can be a surface treatment (chemical or electrolytic). Mating areas must be free of paint, bare metal being generally treated against corrosion. Not all anti-corrosion treatments are good conductors:

- Anodized aluminum is non-conductive.
- Bichromate olive-green, and most aluminum treatments make poor, unstable contacts.
- Alodyne provides a fair conductivity, but the process has been banned due to toxicity, being replaced by neutral chromate treatments, like CHROMITAL/Surtec.
- Zinc, nickel, or cadmium plating provide a good conductivity.

Alternative solution is by conductive tape pressed on bare metal. The contact resistance of such tapes after hard compression must not exceed few mΩ per square. For applications with long-term exposure to harsh environment, beware that the conductive glue backing does suffer with aging, with a trend to polymerization over the years.

With metallized plastic housings, the seam treatment needs only to be proportionate to the box skin SE, which is generally more modest (typ. $\leq$50 dB below 100–200 MHz). If the conductive coating is resistant to abrasion, mating edges can be shaped to provide an electrical continuity, without the need for gasket, for inst. Using tongue-and-groove or other molded profiles for assembly (Fig. 6.15). The flexibility of plastic provides the necessary contact pressure of the conductive surfaces.

6.6.2   Shielding for Cooling Apertures

Several techniques can restore shield integrity at convection or forced-air cooling vents (Fig. 6.16):

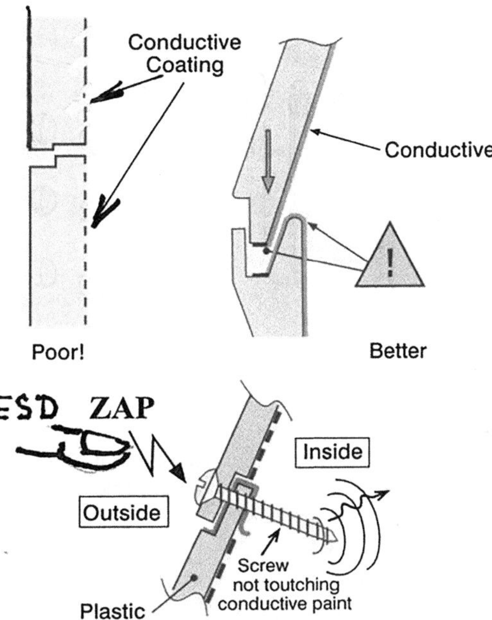

Fig. 6.15 Conductive paint on plastic box could provide an electrical continuity, without the need for gasket. *Notice the mistakes*: on top left, the paint coating is interrupted at the mating edges, causing a breach in the shield. *With the lower mounting*: the screw makes no contact with the conductive paint

(a) Break large openings into several smaller ones, which can be done at virtually no cost if made during stamping or molding of the housing. It may also put the critical circuits at a relative greater distance, compared to the aperture size, reducing proximity effect. Replacing long slots with smaller (preferably round) holes, the SE improves by $\Delta = 20 \log N$, N being the number of new, identical holes. If some depth is added to the wall such that $d/l > 1$, the waveguide term becomes also noticeable, improving SE.

(b) Press a metal mesh over the cooling hole. This screen has to be continuously welded or fitted with a conductive edge gasket having an intrinsic SE superior to the overall objective.

(c) Install a honeycomb air vent if more than 60 dB is required above 500 MHz and up to several GHz, along with a low air pressure drop.

6.6.3   Shielding for Viewing Apertures

Displays, tactile keyboards, meters and the like, being the largest openings on equipment façades, might appear as the greatest contributors to a mediocre SE for a whole enclosure.

On the other hand, compared to the typical RF "hot plate" of a filter mounting or I/O connectors area, high-frequency sources are seldom mounted right on or behind front panels. Tests results show that most equipment can tolerate rather large,

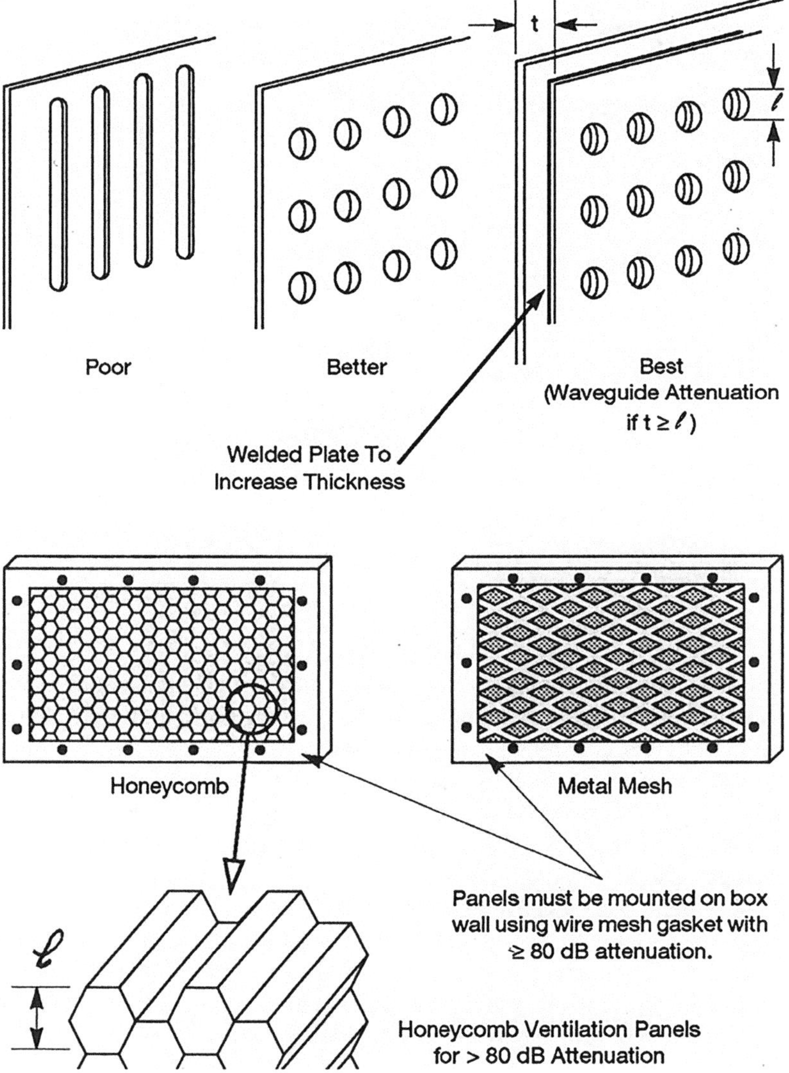

Fig. 6.16 Methods for shielding cooling apertures

unshielded apertures on their user's displays, while ten times smaller slots in the cable entry zone are causing significant leaks. In a sense, the intrinsic SE of any aperture being calculable, its radiation still depends on how it is excited.

Yet, not knowing in advance how RF currents will distribute on the box skin, it is safe assuming that viewing apertures are as prone to leak as any other one. The solutions are:

(a) Fine grid wire mesh, on top of, or sandwiched in the glass, plexiglass or other material. Densities of up to 12 wires/cm (knitted) or up to 100 wires/cm (woven) are available. The performance can be derived from Fig. 6.17. The denser mesh

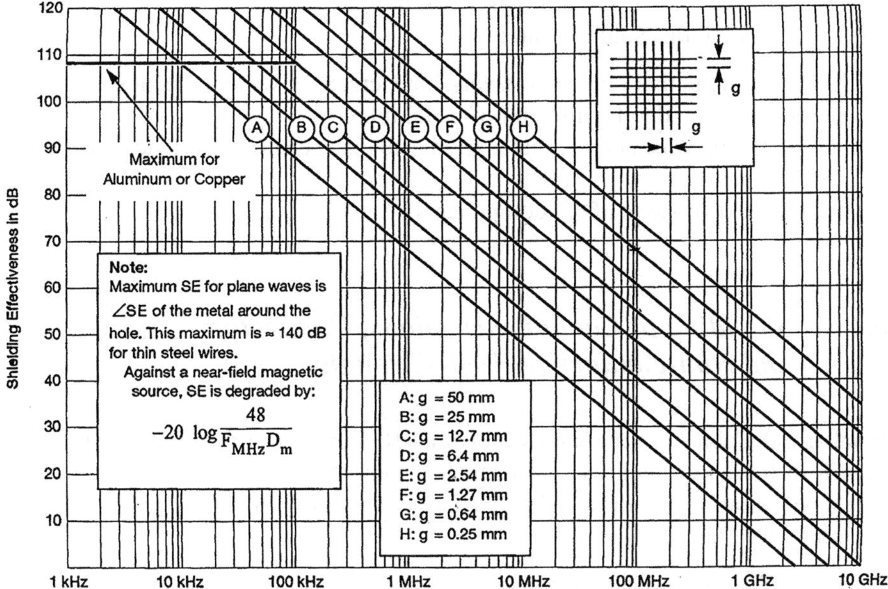

In the note box the degradation factor is:

$$-20 \log \frac{48}{F_{MHz} D_m}$$

Fig. 6.17 SE of wire mesh for cooling or viewing apertures, far-field conditions (distance >48 m/F MHz)

offers more SE, but at the prejudice of transparency. A modern alternative is the litho-photographic deposit of a thin copper screen.

(b) Transparent conductive film, where a thin film of gold or indium tin oxide (ITO) is vacuum deposited on the transparent substrate. Film thickness has to be low (10^{-3} to 10^{-2} μm) to keep an 80% to 60% optical transparency, but thinner film means greater surface resistance. Typical transparent films have surface resistivities ranging from of 50 to 5 Ω/sq., translated in far-field SE of 10–30 dB.

(c) Shielding the display from the rear side: the display is shielded behind the box wall by a doghouse, equipped with feedthrough capacitors for connecting wires.

In all the solutions described above, an EMI gasket is needed at the shield-to-box joint. Often, one such fitting is already provided by the shielded window vendor.

6.6.4   Shielding the Component Holes

Holes for potentiometers shafts, switches, lamps, fuseholders, and so on generally are small. But being present in the middle of metal surfaces that are carrying CM currents from inside the box will enhance the radiation phenomenon. A shaft, lever or fuse will act as a monopole, exiting via a short coaxial line, capable of transmitting radio signals.

As far as FCC, CISPR and other civilian limits are concerned, component holes are seldom a problem because of the relatively small leakage. With tough

MlL-Std-461 or TEMPEST emission limits, these holes can be significant contributors to EMI radiation. Solutions are:

- Use non-conductive shafts or levers, and create a waveguide attenuation by lengthening hole depth with a bit of metal tube.
- Use grounding washers or circular springs to make electrical contact between the shaft and panel.
- Use shielded versions of the components.

6.6.5   Shielding of Cable Penetrations, Connectors and Non-conductive Feed-Throughs

Finally, I/O Cables are the largest potential RF carriers in the entire system. The breach in box skin integrity due to their penetration is a serious concern. Cable shields penetration is indissociable from cable shielding. A cable entry hole is contributing to SE, as follows:

- If an analysis of box SE show that the *cable exit naked hole is tolerable*, the cable can still behave as an antenna. If the cable needs to be shielded for radiation and/ or susceptibility, its shield must connect to the barrier crossing via a 360° clamp, an ultra-short strap or, best of all, a metallic connector shell. If the cable is not shielded but is still a threat, each one of its wires must have been filtered: there is no point in shielding the hole.
- If an analysis of box SE show that the *cable naked hole is not tolerable*, we still must use shielded cable and connectors, re-creating shielded enclosure for the whole interconnect cabling system. A trade-off is to use unshielded cable and to block aperture leakage with a shielded and filtered connector receptacle, acting as a recessed shield barrier behind the cable entry.

Some other exit/entry holes can be found for non-conductive lines such as pressure sensors, fluid lines, fiber optics, etc. If the tube is non-conductive and the SE of the naked hole is not sufficient, this type of leakage is easily reduced by using the waveguide effect. For fiber optics, transmitters, and receivers, metallic packages are available with appropriate tubular fittings.

6.6.6   Aggravated Effect of Box Leakages Near a Cable Penetration

When a cable exit is close to an enclosure seam or slot, the attenuation may appear locally less than its theoretical far-field value. In emission situations, the exciting source inside can couple to the first centimeters of the outer cable segment (Fig. 6.18) by a kind of magnetic or capacitive crosstalk. Using a current probe, CM currents could be found on the cable, even after installing I/O filters or ferrites, turning the

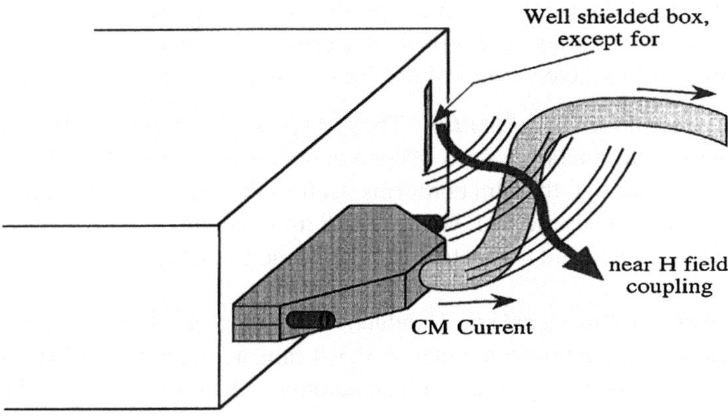

Fig. 6.18 Excitation of I/O cables by a nearby slot. (From [11])

cable into a secondary antenna. Such leakages in a "hot plate" zone must be seriously controlled, for instance, by special care of the enclosure tightness close to cables I/O ports.

6.7 Specially Hardened Equipment Housings

Manufacturers of ready-to-use racks and cabinets offer EMC-treated versions. Even a standard steel or aluminum cabinet with some simple precautions (paint-free and zinc- or tin-plated contact areas, metal mesh air filters) provides some degree of shielding. When fitted with EMI gaskets, shielded air vents, 100% welded frame joints and piano-hinged doors for tight seam tolerances, they offer SE >60 dB up to 150 MHz, 40 dB at 500 MHz, at a cost increase of $350 to $500 (2017 prices) compared to standard version.

A word of caution when dealing with RF emission problems: some SE data shown on shielded cabinet ads are measured per MIL-Std-285 method with a radiating source outside, 30 cm from the doors. This may not replicate cases where the source is inside: proximity effects can cause lower SE than expected, especially when inner devices and cables are near the cover seams.

6.8 Shielding Components for Mass Production, Consumer Products

Technical evolution has brought a huge number of miniature, popular devices using high-speed digital circuits and wireless RF techniques, operating at >1 GHz, urging the development of new shielding hardware. These shielding items have to be economical, lend themselves to mass-production techniques like a production of 1000 or more devices/day, and provide performances which were barely attainable by the costly military electronics of the 1980s. Such SE hardware includes:

- ***Heat-formable, shrinkable films:*** They are generally polymer-fiber coated with low fusion point metal mesh (3 M) or a conductive ink grid (G.E. Lexan). When heated after die-cut, the film conforms itself to the 3D shape of the plastic housing that needs to be shielded. With 0.2–0.8 mm thickness, foil resistance is in the 0.1–1 Ω/sq. range, and the textured nature of the content prevents resonant cavity effects.
- ***Thin, form-in-place gaskets:*** Conductive caulking can be applied in a regular cord-gasket with diameter as small as 0.3–1 mm, acting both as EMI and weather gasket. They can be deposited with an automatic dispenser, or printed in a single operation like an ink, conforming to intricate shapes.
- ***PCB component shields:*** Five-sided cans, stamped from tin-plated steel or brass, are available off-the-shelf in standard shapes, with heights as low as 3 mm. They are wave-soldered to a printed ground belt around the specific component or PCB zone that need to be shielded (Fig. 6.19). The PCB ground plane acts as the sixth side of the enclosure.

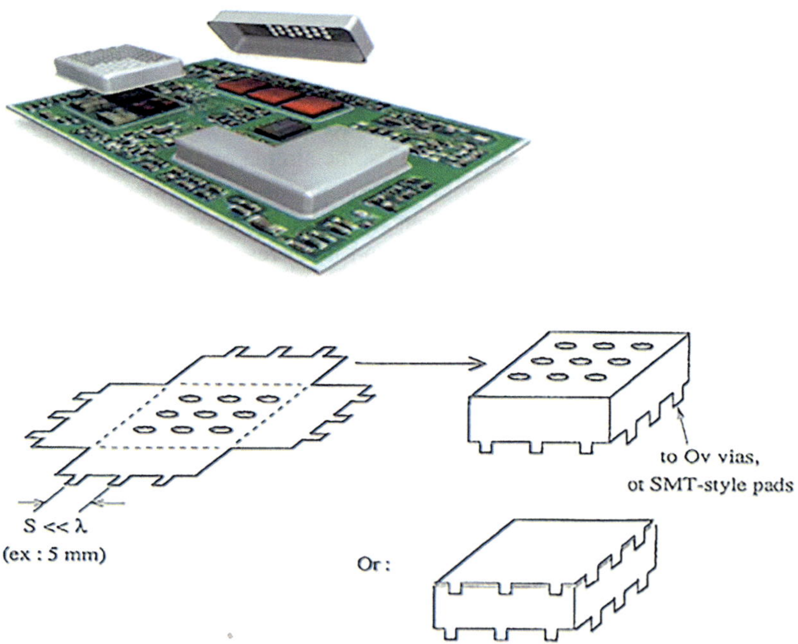

Fig. 6.19 Partial shielding at the IC/module level. (From LAIRD Co.)

6.9 Large Shielded Rooms and Shielded Buildings

Certain applications like EMC testing, medical investigations (NMR) or protection against eavesdropping (Tempest) require rooms, or even entire buildings that are fully shielded with high requirements of 60–100 dB. Such shielding has the same basic aspects as reviewed before, but with additional constraints: the shielded zone must keep all the access and conveniences of ordinary facilities: single or double doors, windows, air conditioning ducts, lighting, and many more that require special hardware elements. Shielded rooms up to 30 m² of floor area are available as self-supporting cages, made of prefabricated galvanized steel wall panels, door frames etc.... that are assembled on site. Larger rooms are shielded by installing copper foil layers on the walls and ceiling, and prefabricated or customized shielded doors and windows.

An other, recent option is to use directly in the basic building a fiber-loaded concrete "Gray-Shield," such as the required SE is already part of the building walls/ceiling, w/o the extra cost of sheet metal or mesh covering of the walls [22].

6.9.1 Summary of Radiated EMI Control Via Box Shielding

1. When the best affordable measures have been taken at PCB and internal wiring level, the equipment housing is the ultimate barrier against radiated emissions.
2. Until the last hole or slot is checked, the best metal box could appear to be useless as a shield.
3. For metal housings:

    (a) Bond all metal parts (a floated item is a candidate for re-radiation).
    (b) Avoid long seams and slots: a 30 cm seam is a total leak at 300 MHz and above.
    (c) Use gaskets or waveguide effect: design fold-over shapes for the cover edges.

4. For plastic housings: use conductive coating ≤ 2 Ω/sq., then treat like a metal housing. Avoid long, protruding screws inside.
5. Maintain or restore shield integrity at: cooling holes, viewing apertures, component/cable penetrations.
6. Beware of noisy circuits or cables close to seams and slots: they degrade an otherwise adequate SE.

Quiz

There is ONE good answer, or ONE that is better than the others.

1. A shield with a SE of 70 dB provides a field attenuation of:

    (a) 300 times
    (b) 7000 times
    (c) 70 times
    (d) 3000 times

2. To be in far-field conditions, a shield must be:

    (a) at more than one wavelength away from the radiating source
    (b) at more than 1/6 wavelength from the source, if the latter is a magnetic antenna
    (c) at more than 1/6 wavelength from the source, whatever the type of antenna
    (d) near or far-field conditions depend if the source is inside or outside

3. A good reflection is obtained when:

    (a) the shield is in far-field conditions
    (b) the wave impedance of the field is $\ll$ metal surface resistance
    (c) the wave impedance of the field is $\gg$ metal surface resistance
    (d) the radiation is from a magnetic source

4. Conductive films and paints work on the principle of:

    (a) reflection and absorption
    (b) mostly reflection
    (c) absorption
    (d) optical resonance

5. An aperture in a shield wall:

    (a) does not degrade the metal barrier SE as long as its own attenuation is not < than metal SE
    (b) always degrade the SE of a metal barrier
    (c) is only leaking if its orientation is // to field polarization
    (d) is only a concern if critical circuits are very close to the aperture

6. To obtain the total SE of an enclosure with apertures one must:

    (a) Add the SE(dB) of all the apertures
    (b) Take the antilog of each aperture, combine their total and recalculate the global SE
    (c) Retain the largest value of SE
    (d) Add the SE(dB) of all the apertures and substract from the SE of original metal barrier

    YOU CAN FIND THE CORRECT ANSWER AT page xxx.

Chapter 7
Coupling from and to the Power Mains, the Fifth Coupling Path

Preamble

Former chapters reviewed four of the five principal conduction and radiation coupling mechanisms, as they affect equipment/system susceptibility and emissions. The last ones were addressing the shielding of equipment boxes, from the smaller hand-held devices up to large cabinets or even entire rooms. This chapter is covering the fifth, very important coupling mechanism: how the various disturbances that exist on the power mains (public AC distribution, ship or aircraft AC distribution, vehicle dc distribution, and so on) can find their way from their source down to the electronic components of a system? Reciprocally, some active elements in our equipment (dc–dc and ac–dc switch mode regulators, fast IGBT and thyristors, variable speed drives, fast logic circuits) can in turn become EMI sources, polluting the power distribution to the prejudice of other users. And like for the other coupling paths, many solutions to power line susceptibility problems will also reduce conducted emissions.

7.1 Equivalent Circuit for Typical Power Distributions

Whatever the installation, you must know at least approximately what is the wiring and grounding scheme of your power distribution, from the power source down to the users. This knowledge is important for:

- understanding and estimating the coupling factors
- understanding the common mode (CM) to differential mode (DM) conversion
- selecting the most appropriate, and economical filter
- selecting the proper characteristics for surge protection devices.

Figure 7.1 shows the most common power distribution schemes. Depending on the applications, a power distribution can be fairly simple: in most vehicles the

© The Author(s), under exclusive license to Springer Nature
Switzerland AG 2025
M. Mardiguian, *ElectroMagnetic Compatibility*,
https://doi.org/10.1007/978-3-032-02688-0_7

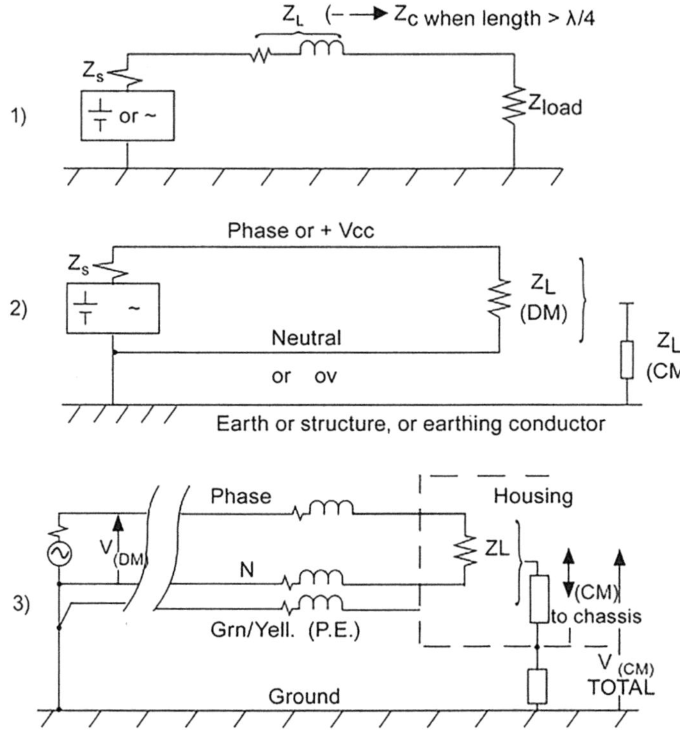

Fig. 7.1 Most common configurations for power distribution

battery (+) is running in the entire harnessing, the (−) return conductor being the metallic car body. Otherwise, it is generally more complex with the (+)/(−), or Phase/Neutral, running together in the entire power distribution network. In houses and professional buildings, the Ph and N wires are run along with a third wire, acting as a safety lifeline to the earthing electrode(s), where the neutral of the power source (entry transformer or power plant) is also grounded. Therefore as shown on the schematic, we have in general:

(a) a Diff. Mode impedance between the hot wire and its return.
(b) a Com. Mode impedance between either one (or both) power wires and the local structure ground or chassis.
(c) If a there is safety earth (green/yellow) wire, two values for CM impedance can be considered:

- The Ph+N wires vs. the earthing wire, which is only valid as long as the earthing wire is sufficiently short and the frequency low enough for considering this wire as the earth potential.
- The Ph+N+ Earth wires altogether vs. the real local ground, which can be a structural ground, concrete slab, ship hull, and many more. In this later case, the CM currents will flow in the Power + Earthing wires in the same direction, returning to the Power source via the local ground reference.

7.2 Power Mains Impedance

This is what an equipment will see when looking toward its power source. Figure 7.2 shows the general impedance profile of urban power distribution, made after compilation of wide statistical surveys in major cities, worldwide. The thin dotted plots are the upper and lower 10% of the measured impedances (that is less than 10% of the measured data points were above or below these values). Shown for comparison is the profile of a Line Impedance Stabilization Network (LISN) that is deemed to represent the average impedance value of urban power mains (See last chapter "Military & Civilian EMC Norms").

Both CM and DM Power Mains impedances must be considered, with values depending on the application: public network, vehicle, aircraft, and ship, which dictate a typical length of wiring from the source to the users, and the Neutral, or return conductor grounding practices. In any case, the overall shape of Power Mains impedances is the same, when seen from the end user Power Entry terminals, or simply the wall outlet:

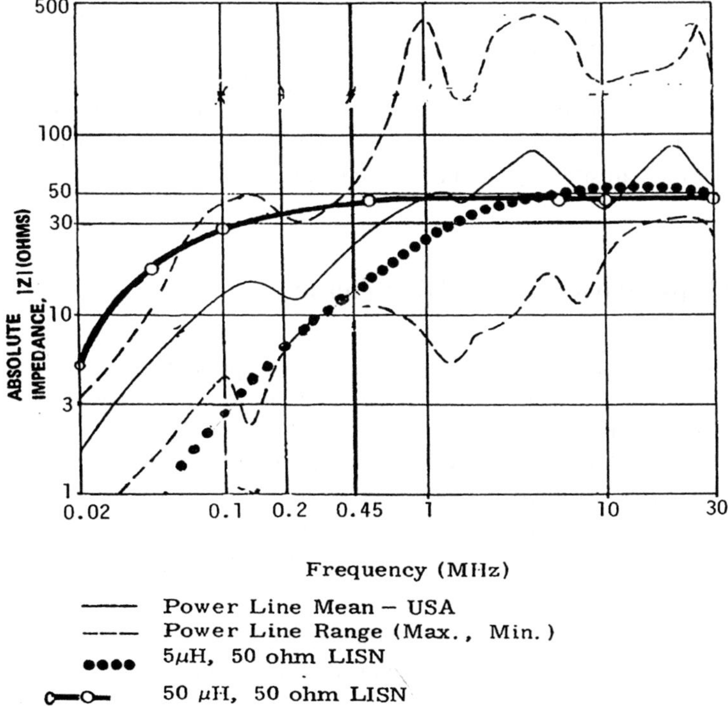

Fig. 7.2 Power lines and artificial network (LISN) impedance profiles. The 5 μH/50 Ω LISN is used for EMC test of motor vehicles and civilian aircrafts

- From dc to a few kilohertz, it is basically the wires ohmic resistance, plus the source resistance, this latter being obtainable from the alternator or substation transformer short-circuit current.
- From a few kHz to $\approx$MHz, the impedance increases linearly with frequency, since it is dictated by the self-inductance of the mains wires, the wide spread of values being due to the overhead or buried type of wiring, the proximity of a ground plane, the number of users in parallel, etc.
- Above the megahertz region, the impedance stabilizes around a mean value of 50 Ω for CM (line-to-ground) and 100 Ω for DM (line-to-line), corresponding to the line characteristic impedance.

7.3 Equipment Input Impedance

Knowing the equipment input impedance is important too, since it affects the behavior of the equipment to EMI coming in, as well as filters and surge suppressors selection. Without EMI filter, the impedance at input terminals is that of the front-end transformer (often the first regulator), with its secondary load transferred to the primary. If there is no front-end transformer, the power input being dc or rectified ac, the DM input impedance is low, being that of the direct-polarized rectifiers and associated buffer capacitor. When the power input is isolated from the equipment frame or signal ground, as often the case with 230 V/50 Hz power, the CM input impedance is high at low frequency, decreasing progressively when F increases, because of the parasitic capacitance of the floated circuits.

7.4 Major Power Mains Disturbances

Power line disturbances are a frequent cause of equipment malfunction or even damage. They can be quasi continuous like HF noise superimposed to the normal mains voltage, waveform distorsion and notches, or incidental, like transient overvoltages, spikes, short partial or total voltage drop (Fig. 7.3). Most are not due to the quality of the power source, but to the operation of the various users, whose current demands are often irregular or pulsed, causing voltage dips or surges on the distribution. Lightning strokes on, or nearby the power lines are also a frequent cause of severe overvoltages.

- Slow or steady fluctuations of line voltage can reach $-/+$ 10%, regarded as normal. All modern equipment are equipped with regulators that can easily handle such variations.
- Flickering: periodic, very low frequency undervoltage, causing lamps blinking.

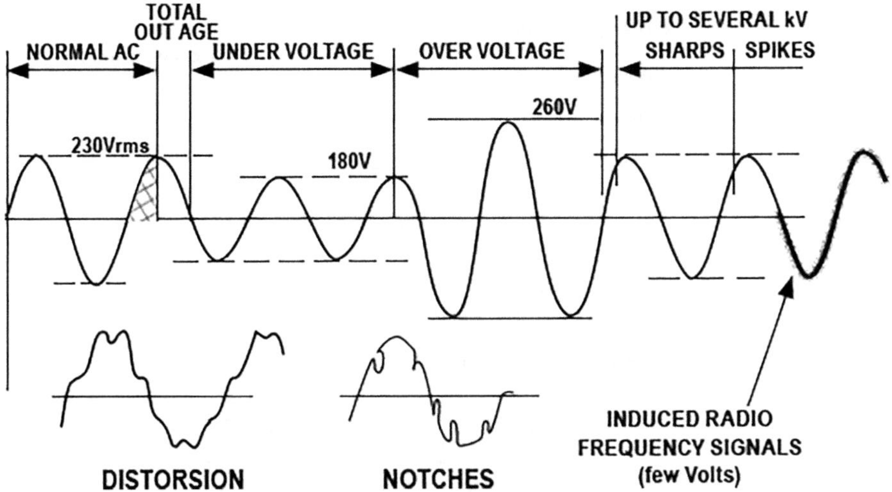

Fig. 7.3 Major power line disturbances

- Ripple or HF noise riding over the supply voltage can be caused by poor filtering of switch mode regulators, or strong ambient RF field coupling onto the power lines. These rarely exceed a few volts and are easily reduced by EMI filtering.
- Harmonic distorsion and "notches" in the AC voltage sinewave are caused by currents peak drawn by non-linear loads of the various users.
- Unipolar or ringing transient overvoltages and spikes can often exceed hundred Volts, superimposed to the AC (or dc) voltage. Extreme values of 4–6 kV can be reached with severe lightning indirect effects or Fast Switching transients. These can be damaging and require transient suppressors.

7.5 Improving Equipment Immunity to Power Line Disturbances

Several variables are considered when hardening an equipment or an entire system (Figs. 7.4 and 7.5):

- The nature of the disturbance: under/overvoltages, exceeding the normal +/− 10% fluctuations, energetic high voltage transients, short spikes, total outage?
- The duration of the disturbance: ms? s? more?
- The risk assessment vs. the probability of occurring: how many times a day, week, year?
- The DM or CM nature of the incoming disturbance, as seen by the victim equipment.

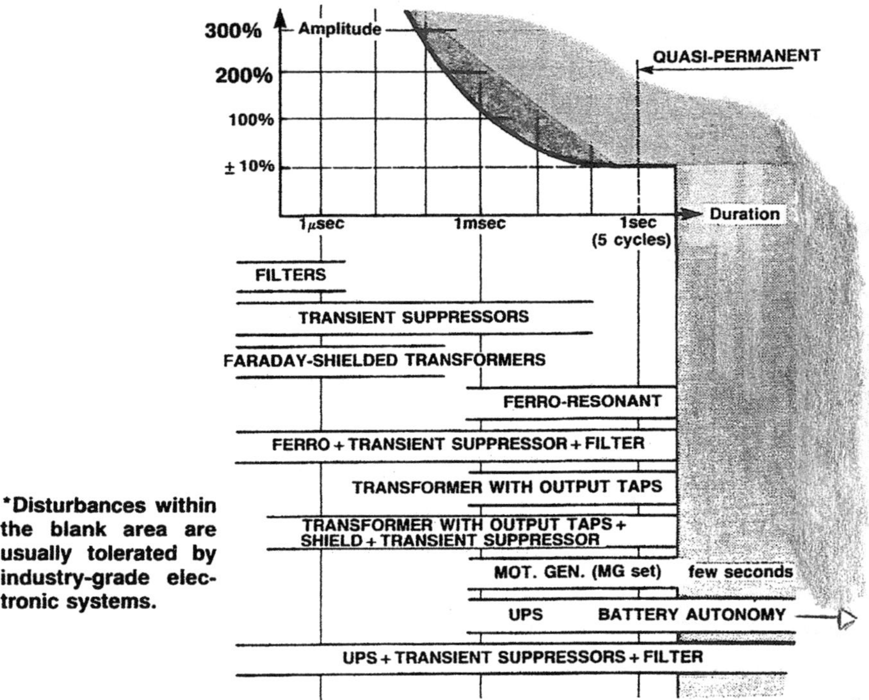

Fig. 7.4 Classification of power line disturbances according to their severity/duration and related solutions

7.5.1 *Filters for Reducing Powerline Interference*

Practically no modern equipment, with its fast digital circuits, switch mode power supply regulators and eventually RF devices, could meet EMC requirements without an efficient filtering. Yet, EMI filters are too often chosen on an empirical basis or a vague belief in manufacturers catalog performances. A good EMI filter must:

(a) attenuate the incoming power line noise to make the equipment immune to the most severe DM and CM aggressions expected in its normal environment,

(b) attenuate the HF noise caused by the equipment to its own power input, both DM and CM, such as it does not violate conducted EMI emissions limits.

Both (a) and (b) performances must be achieved in a system whose source and load DM and CM impedances are poorly defined. On top of these challenges, the filter must comply with size, weight, and cost limitations, as well as constraints on maximum permitted values for line-to-ground capacitance and high voltage withstanding tests. Needless saying, a filter is generally the result of a tight trade-off.

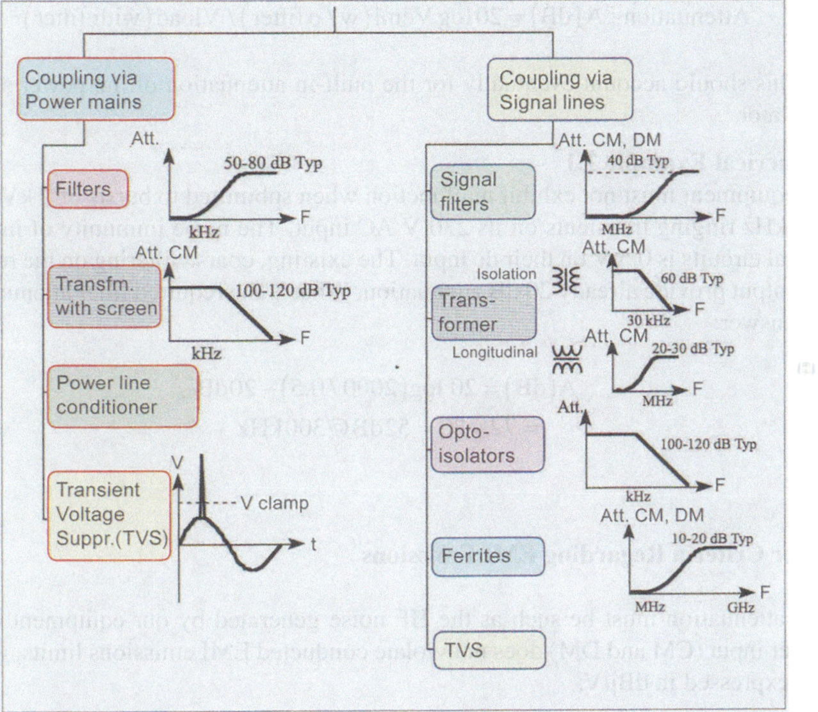

Fig. 7.5 Overview of conducted EMC solutions to permanent or transient EMI: left branch is for power lines disturbances

7.5.2   Selecting a Filter

Given that an EMI filter is a low-pass element, the following parameters dictate our choice: the required attenuation (DM/CM) vs. frequency, the filter cut-off frequency and its number of poles, the impedances on the source and load sides of the filter, that in turn will guide the filter scheme: capacitive-only, inductor-only, L-C, Tee or Pi type?

The normal service voltage/current must also be considered for the filter elements.

Filter Criteria Regarding EMI Immunity

The attenuation must be such as the most susceptible circuits inside the equipment, must not receive on their dc input a noise voltage greater than their threshold of sensitivity.

$$\text{Attenuation}: A\,(\text{dB}) = 20\log \text{Vemi}\,(\text{w}\,/\,\text{o\,filter})\,/\,\text{Vload}\,(\text{with filter})$$

This should account eventually for the built-in attenuation of the power supply regulator.

Numerical Example 7.1

An equipment must not exhibit malfunction when submitted to bursts of 2 kV(CM) 300 kHz ringing transients on its 230 V AC input. The noise immunity of internal digital circuits is 0.5 V on their dc input. The existing, coarse filtering on the regulator output provide already 20 dB attenuation. What is the required filter attenuation?

Answer:

$$A(\text{dB}) = 20\,\log(2000\,/\,0.5) - 20\text{dB}$$
$$= 72 - 20 = 52\text{dB}\,@\,300\,\text{kHz}$$

Filter Criteria Regarding EMI Emissions

The attenuation must be such as the HF noise generated by our equipment on its power input (CM and DM) does not violate conducted EMI emissions limits, generally expressed in dBμV:

$$\text{Attenuation}: A\,(\text{dB}) = \text{Vemi}\,(\text{dB}\mu\text{V}, \text{w}\,/\,\text{o\,filter}) - \text{VdB}\mu\text{V}\,(\text{spec.limit})$$

Example 7.2

Early testing on a prototype has shown spurious harmonics of the switch mode power supply with amplitudes of 104 dBμV from 350 kHz up to 3 MHz. The EN 55022 class B limit is:

- 350 kHz: 50 dBμV
- 1 MHz: 46 dBμV.

What is the required filter attenuation?
Answer:

- 350 kHz: A dB = 104 dBμV − 50 dBμV = 54 dB
- 1 MHz: A dB = 104 dBμV − 46 dBμV = 58 dB

  * Remarks:

- When subtracting dBμV from dBμV, result is in dB, since arithmetically we are making a ratio of two voltages.
- the filter requirement can be based on 54 dB at 350 kHz, which will also satisfy the 58 dB goal at 1 MHz (filter attenuation normally increases with frequency).

7.5.3   Number of Elements in a Filter

The number of poles in a filter (that is the number of L, C elements being cascaded) is determined by the desired attenuation at frequency F_{emi}, and by F_{co}, the cut-off frequency (or -3 dB point), below which the filter has no attenuation. Generally, the cut-off frequency of power line line filters for individual equipment has F_{co} values in the 10–30 kHz range. Figure 7.6 shows the attenuation of any filter, given its cut-off F_{co} and number of poles (n).

Numerical Example 7.3
With the filter of Example 7.2, we were looking for 54 dB at 350 kHz. Assuming a cut-off at 10 kHz, how many poles our filter should have?
    Answer:
    From Fig. 7.6, we see that, given $F_{emi}/F_{co} = 350$ kHz/10 kHz $= 35$, the filter must be at least a second order (n = 2).

7.5.4   Influence of Actual Source and Load Impedances

Filter manufacturer's data are generally measured in a 50 Ω/50 Ω set-up (per Mil Std 222), while actual in situ attenuation may differ significantly. It is strongly dependent on the value of the impedances seen on both sides, to such extent that the performance shown on the vendor data sheet is probably the only one you will never get.

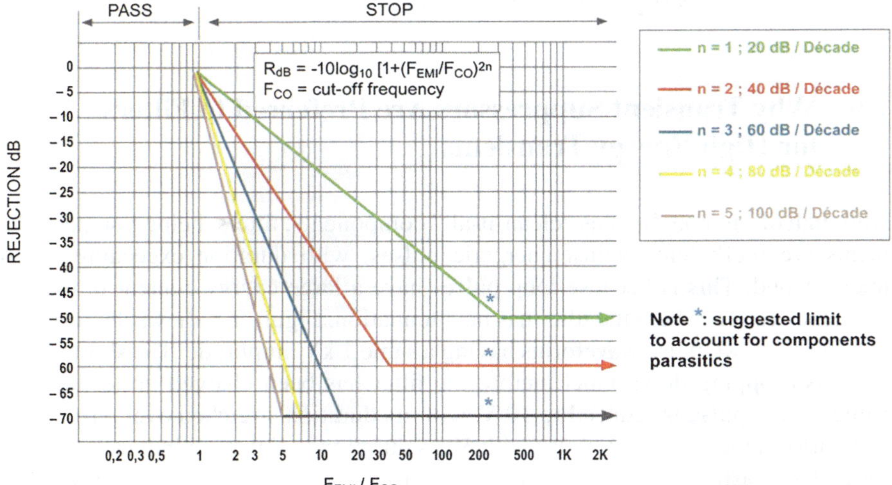

Fig. 7.6 Attenuation of low-pass filters normalized to the F_{emi}/F_{co} ratio for several values of "n"

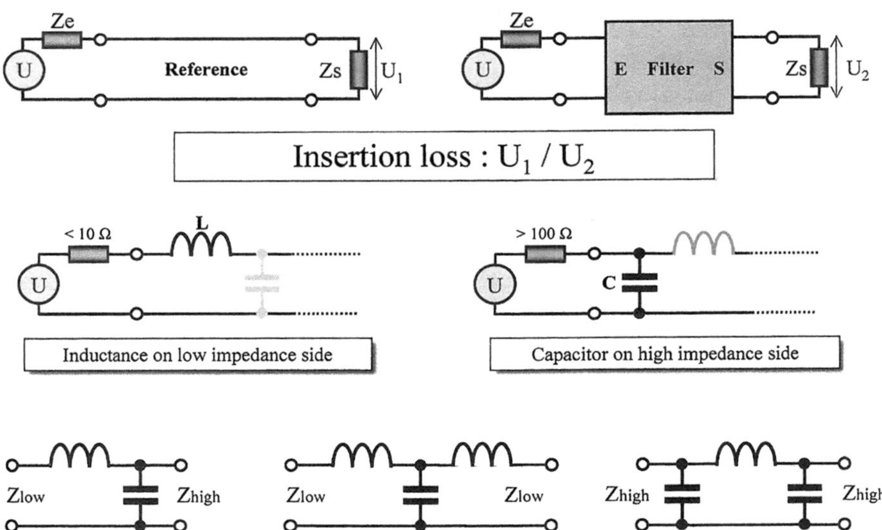

Fig. 7.7 Recommended choice for filter L, C structures, depending on the input/output impedances

Figure 7.7 shows the recommended choice for filter structure, depending on the up and downstream impedances. More on this subject will be found in next Chap. 8, devoted especially to EMC filters, but even without elaborate calculations, the following, simple rule can dictate a preliminary choice for the filter structure:

With filters, *capacitors should look toward high impedances both sides, inductances should look toward low impedances*, both sides.

7.6 Why Transient Suppressors Are Preferred to Filters for High Energy Transients

Although often thought of as "clean-it-all" components, filters, being low-pass elements are inefficient against energetic pulses, with duration exceeding a few microsecond. This is because long pulses have a large energy content in the low frequency range, where the filter has no attenuation. Figure 7.8 shows the response of a filter for two pulse waveforms having a same 1 kV amplitude. The short 100 ns pulse is dropping down much before the filter has been charged up to the peak value, so the pulse is reduced to 10 V, with its duration stretched over >10 μs. By comparison, the same 1 kV pulse with 50 μs duration (a lightning induced transient for inst.) is lasting long enough for the filter to get charged up to the peak value. What is needed is a *component that is not frequency-selective but amplitude selective*, as seen next.

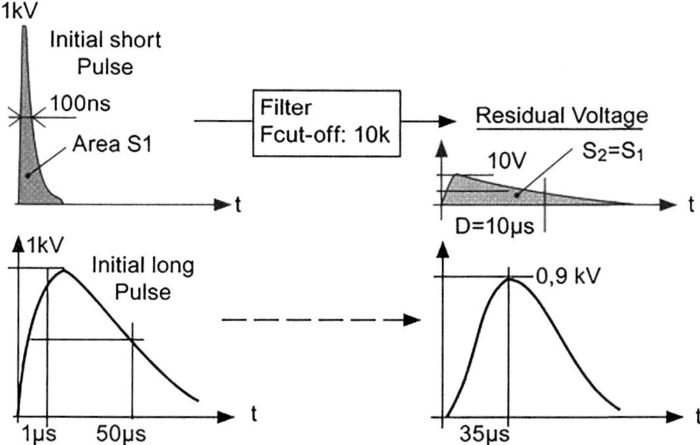

Fig. 7.8 Disappointing results of a power line filter against a long duration pulse

7.7 Transient Voltage Suppressors

Suppressing high voltage, energetic surges is the role of Transient Voltage Suppressors (TVS) like Varistors, Transzorbs or Gas tubes. These non-linear devices are an open circuit up to a breakdown value, above which they behave abruptly as shunting elements with large current capacity. We will review the three principal types of TVS: MOV, TransZorb and Gas Tube.

7.7.1 Metal Oxyde Varistors (MOV)

They are avalanche devices, with an "Off" resistance $\geq 10^6$ Ω. When the applied voltage reaches the V_{br} avalanche threshold (Fig. 7.9), their "ON" resistance drops to a few Ohms. Their main features are:

Advantages
- fast response, in the ns range
- low cost
- fair energy handling (20 J for a 14 mm device)

Drawbacks
- they turn into open circuit destructive mode when I_{max} is exceeded: can be a drawback, or an advantage, depending on the application,
- slanted $\Delta I/\Delta$ slope, causing the actual clamping voltage for I_{max} to be two or three times the V_{break},
- large parasitic capacitance (a few nF) that must be taken into account.

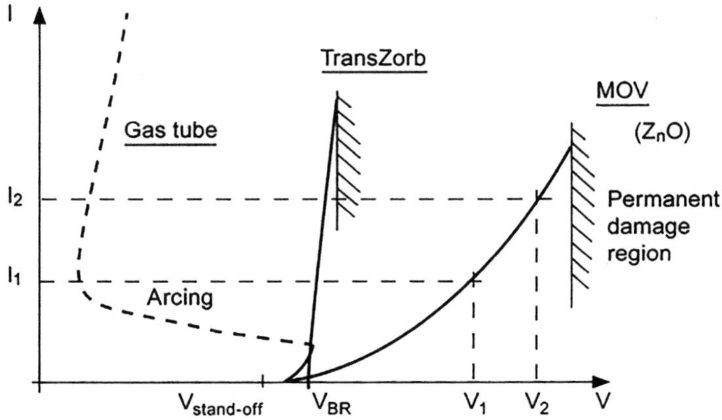

Fig. 7.9 Current/voltage characteristics of MOV, TransZorb® and gas discharge tubes

7.7.2 Zener Diode-Based TVS (Transzorb®, Transil®)

These are special Zener diodes with enhanced power dissipation capability. Their V_{br} threshold is more accurate, with a much sharper $\Delta I/\Delta V$ characteristic.

Advantages
- fast response <ns
- almost vertical $\Delta I/\Delta V$ slope, thus a tight clamping voltage at I_{max}, typ $\leq 1.5 \times V_{break}$
- fair energy handling

Drawbacks
- They turn into short-circuit destructive mode when I_{max} is exceeded: can be a drawback, or an advantage, depending on the application
- large parasitic capacitance (a few nF) that must be taken into account

7.7.3 Gas Tubes

Gas tubes are a modern version of the air gap arrestor that was based on the arcing of a calibrated gap when the air breakdown voltage is exceeded. Modern gas tubes consist in a sealed envelope filled with low pressure gas and a precise interval between the electrodes.

Advantages
- high current handling capability since the terminals are practically shorted during the arcing
- a low parasitic capacitance (a few nF) that must be taken into account

Drawbacks
- slow to react, with delay $>\mu s$
– the line in short-circuit when arcing, until the voltage waveform crosses zero
– not to be used on dc power lines, because the arc will be sustained by the dc voltage and will not extinguish, until a fuse blows, requiring a manual action for re-powering the equipment
– does not exist for triggering voltages below ≈ 90 V

"Solid-State gas tubes": Solid State versions of the spark gap exist in form of thyristor-based crowbar devices that turns on when a certain voltage threshold is reached, putting the line in short-circuit, protecting the equipment components and eventually triggering a breaker or fuse.

7.7.4   Selecting the Proper Surge Suppressor

The selection steps are summarized below:

– Select a triggering voltage (V_{br}) $\approx 10\%$–20% above the highest peak voltage of the power line.
– Estimate the maximum peak current of the transient pulse. It can be found from the transient immunity specification applicable to the equipment. A better figure can be derived from the I, V load curve of the test generator, after subtracting the clamp voltage from the generator open voltage.
– Check this peak against the maximum, no-damage peak current for the selected TVS.
– Check that the peak power: ($V_{clamp} \times I_{max}$) is tolerable for the applied pulse duration.

7.8   Isolation Transformers

An isolation transformer has distinct primary and secondary windings, breaking the CM ground loop between the external power lines and the equipment or system. The primary-to-secondary magnetic coupling allows the normal transfer of the DM (line-to-line) voltage, while the galvanic isolation prevents the CM voltages from passing through the transformer. An ordinary isolation transformer is an efficient barrier for opening low frequency ground loops between an equipment (or an entire installation) and the power mains.

Yet, as frequency increases, the high value of the barrier isolation (typically $>10^6\ \Omega$) becomes shunted by the inter-winding capacitance. This capacitance is ranging from 50 to 100 pF for small equipment transformer up to several nF for transformer in the kVA range, causing the primary/secondary protection against loop currents to almost disappear around a few megahertz. Special winding sequence and more physical separation between the winding stacks can improve the CM rejection by about 10 dB.

7.8.1 Faraday Shielded Isolation Transformers

For more HF isolation, the best technique is to use a Faraday-shielded transformer [2]. A non-closed foil is wound between the primary and secondary windings and grounded locally, such as a CM voltage appearing on either side is stopped by the shield, and the capacitive current is sunk to the chassis or structural ground (Fig. 7.10). An other advantage of an isolation transformer is to allow recreating locally an earthed Neutral on the secondary side in a specific zone of a large facility.

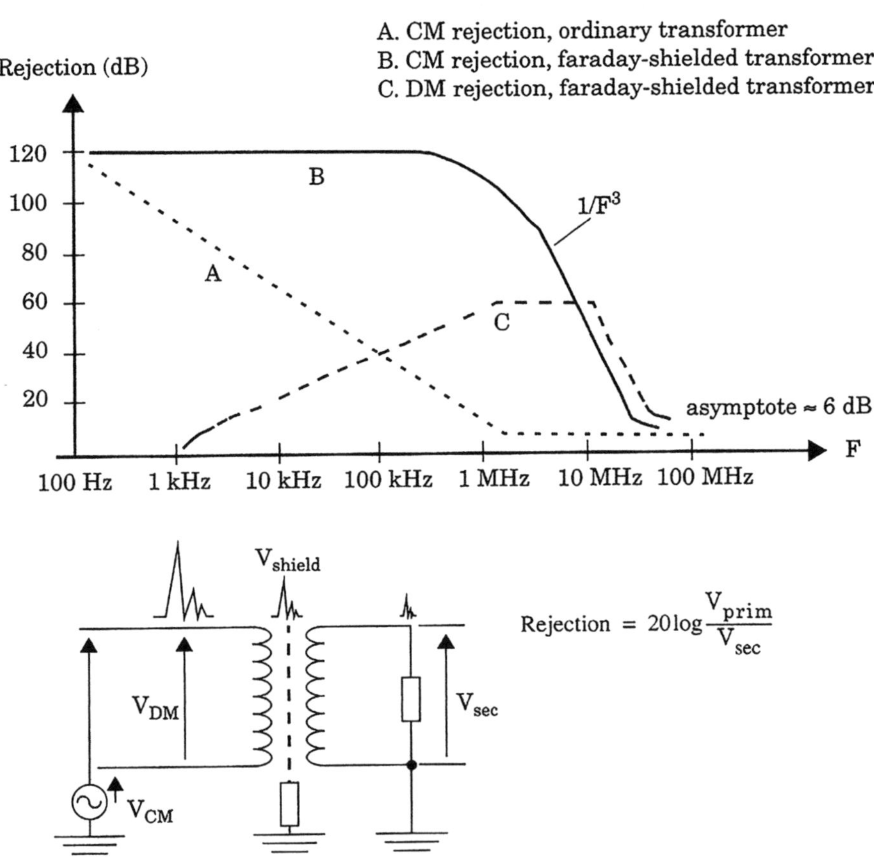

Fig. 7.10 EMI Rejection of ordinary and Faraday shielded Isolation transformers. Notice the small parasitic impedance of the shield ground connection that could cause some re-injection of noise in the secondary. (From [13])

7.9 Back-Up and Uninterruptible Power Supplies

When the problems are caused by serious disturbances, or total outage of the normal ac mains, the ultimate resource is to locally re-create a more dependable power delivery, using line conditioners or uninterruptible power supplies (UPSs).

A *line conditioner* is a special transformer with an automatic regulation circuit on its secondary. For power line over or under voltages like +/−20%, these devices can provide a secondary voltage within 5% of the nominal value. With modern electronic equipment using switch mode power supplies, this feature has less interest because switcher can keep a constant dc output even with primary voltage varying within 100–250 V. Yet, there are still many electrical equipment (motors, machine-tools, lighting appliances) whose longevity rely on an ac supply within 5% tolerances. But, except for ferro-resonant transformers that can keep-up with one missing cycle (10 ms for 50 Hz ac), a line conditioner has no energy storage and cannot make-up for a long 50% voltage drop or a total outage.

A *UPS* is an ac/dc + dc/ac converter coupled with a battery bank. It can provide a well-regulated ac output, even during a momentary, full interruption of the regular ac mains. Two basic principles are used: "Off-line" (or standby) and "On-line." For both devices, the autonomy is depending on the size, in Amp-hours of the battery storage.

With the Off-line UPS type, the loads are normally fed by the regular ac supply, while the ac/dc converter is keeping the battery bank fully charged. In case of ac mains shortage, the ac distribution to the secured loads is automatically switched on the dc/ac inverter, until the regular ac mains return to normal. Notice that this system provides no protection to HF disturbances and transients of the normal ac mains. It is the simplest system, typically used for small installations, like small offices, shops, and so on, where most of the sensitive equipment like appliances, PCs and the like are normally equipped with EMC protection.

With the On-line type, the secure loads are always supplied from the dc-ac inverter, hence decoupled from the regular ac mains, which serve only to charge the battery bank. In case of inverter failure, a no-interruption static switch is transferring the privileged load line to the regular ac. This static switch is, in turn, a possible cause of EMI problems because the solid-state switches are a mediocre isolation barrier against high frequency. This can be avoided by careful EMC filtering and transient protection of the back-up line.

7.10 Emission Aspects with Power Line Coupling

One earliest cause of user-created disturbances to the power mains have been the ac–dc rectification and gate-controlled rectifiers used in many equipment. The increasing use of high frequency converters like switch mode power supplies and inverters, variable speed drives, light dimmers, fluorescent lights, has shifted the

noise emissions toward higher frequencies, causing both conducted *and* radiated
EMI issues. In this respect, EMC filters should be regarded not only as a cure against
conducted interference up to $\approx$30 MHz, but also against radiated emissions from the
power cord at frequencies >30 MHz.

7.10.1 Poor Power Factor and Sinewave Distorsion Caused by Non-linear Loads

With a half-wave ac rectification like on Fig. 7.11, the current Ic charging the tank
capacitor C is not a sine wave: it is a short current peak whose the product Ic × duration (that is Coulombs) is in turn, delivered to the load for a minimum ΔVr ripple
on the dc output. Since power mains is not a zero impedance source, this brief, but
high amplitude current is causing a small voltage dip on the top of the V_0 ac
sinewave.

Similar phenomena occurs with full-wave rectification or three-Phase rectification (six diodes or six gate-driven thyristors). All these schemes are causing
"notches" in the ac mains voltage waveforms. Distorsions appear as harmonics,
spoiling the power factor (the ratio of the fundamental current to the rms addition of
all the harmonic terms I_1, I_2, I_3, etc. …$\rightarrow$ I_n). In civilian applications, maximum distorsion is ruled by IEC 555-2 norm, which require a control of the power factor by

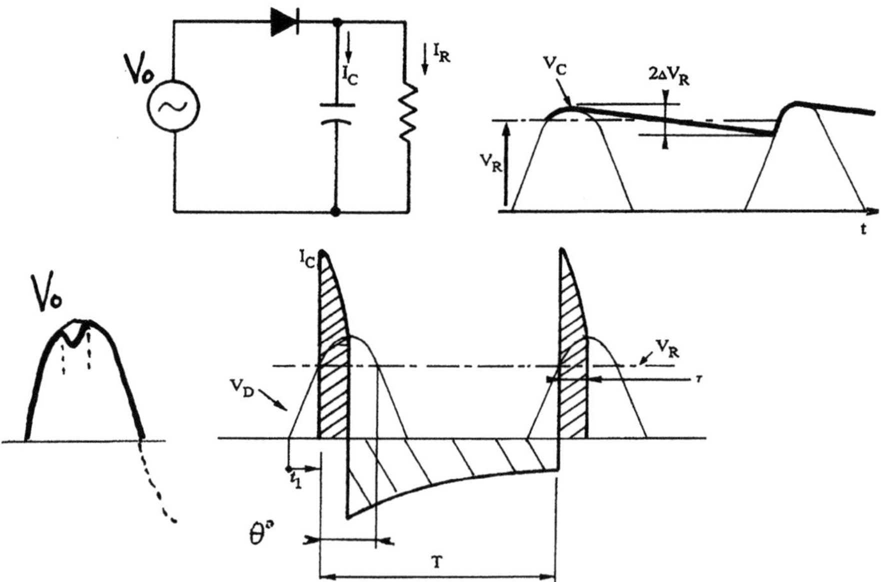

Fig. 7.11 Voltage waveform distorsion caused by the peak current demand of the post-rectification
capacitor

the end users of the power utility. The solutions consist in adding inductances in front of the capacitor, smoothing the rise time of the current demand, or using a Power Factor Controller (PFC), which is an active regulator acting as a dynamic current source.

7.10.2   HF Emissions Caused by Switch Mode Power Supplies and Inverters

Switchers are the main cause of conducted emissions, violating the Military and Civilian RFI limits by as much as 40 or 50 dB for unfiltered items. Several mechanisms, internal to the switcher, are contributing to these emissions (Fig. 7.12):

– *The CM conducted emission path* is generally the major limits violator, for equipment fed by 115 or 230 V ac. The leakage current is caused by the stray capacitance (2) of the switching transistor, IGBT or MOSFET to their heatsink or nearby chassis. Leakage current (Icm) closes by the chassis ground, returning via the power mains impedance then back to the equipment phase and neutral input wiring. For testing, noise level is measured at an artificial mains (LISN) socket.

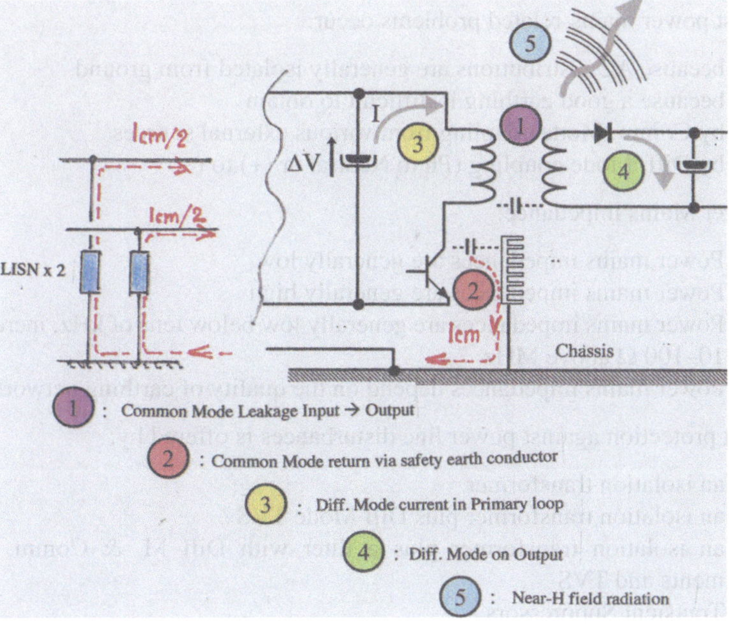

Fig. 7.12 The various mechanisms participating HF noise emissions from a switch Mode Power Supply. The switching transistor voltage is injecting CM currents (2) in the primary line-to-earth loop. The resulting voltage is measured on the artificial power mains networks (LISN) on the left

– Next to the CM current, the DM current (3) circulating in the power input loop is due to the switcher input capacitor bank. Since this capacitor has parasitic series resistance (ESR) and self-inductances (ESL), the flow of the main switching current creates a differential voltage ΔV across the primary input.
– (4) is the ripple and noise on the regulated dc output, caused by fast switching on the primary.
– (5) is the strong local H field radiated b the transformer and chokes windings.

List of Solutions for Reducing Conducted Emissions of Switch Mode Power Supply

• CM current to Ground $\longrightarrow$ Use electrostatic screens, Use CM filter on input.
• CM noise on output dc loads $\longrightarrow$ Use transformer with screen, Apply special precautions in winding the transformer, Use CM filter on dc output.
• DM Noise $\longrightarrow$ Use low ESR/ESL capacitors $\longrightarrow$ DM filters.
• Recovery Spikes $\longrightarrow$ Use fast recovery devices $\longrightarrow$ Install RC snubbers.

More on Switch Mode Power Supplies EMI emissions will be described on next Chap. 8.

Quiz

There is ONE good answer, or ONE that is better than the others.

1. Most power mains-related problems occur:

   (a) because AC distributions are generally isolated from ground
   (b) because a good earthing is difficult to obtain
   (c) by Comm. Mode coupling from various external sources
   (d) by Diff. Mode coupling (Ph to Neutral or (+) to (−))

2. Power Mains impedance:

   (a) Power mains impedances are generally low
   (b) Power mains impedances are generally high
   (c) Power mains impedances are generally low below tens of kHz, increasing to 10–100 Ω above MHz
   (d) Power mains impedances depend on the quality of earthing network

3. Best protection against power line disturbances is offered by:

   (a) an isolation transformer
   (b) an isolation transformer plus Diff Mode TVS
   (c) an isolation transformer plus a filter with Diff M. & Comm. M. elements and TVS
   (d) Transient Suppressors

4. Transient Voltage Suppressors (TVS)

   (a) protect against high-frequency transients
   (b) protect against short and long duration transients
   (c) eliminate RF noise riding over power line voltage
   (d) offer total protection against overvoltages as well as undervoltages

5. Against long duration (>0.1 s) disturbances and RF noise, one should use:

   (a) a UPS plus TVS and EMC filter
   (b) a power line EMC filter
   (c) an isolation transformer
   (d) Diff. Mode and Comm. Mode TVS

Chapter 8
EMC Filters: Design, Selection and Mounting of Power and Signal Lines Filters

Preamble

This chapter goes deeper into one of the simplest, most compact, and economical piece of the entire EMC arsenal: the filter. With current handling ranging from tens of mAmp for signal filters up to more than hundreds Amps for power line filters, they exist in all sorts of size, volume, and package. They can be optimized against common mode (CM), differential mode (DM) interference, or both.

No matter if you look for a commercial, off-the shelf (COTS) or make your own, you must understand how they work and how to determine the most adequate arrangement of L, C elements. A filter should not be picked-up off an EMC guru's tool case: it must be designed. Finally, although regarded as a cure against conducted EMI, filters can also reduce radiated interference, for Emission or Susceptibility (Fig. 8.1).

8.1 Selecting the Right Filter

Given that an EMI filter is a low-pass element, the following parameters dictate its choice:

(a) the required attenuation (DM/CM) for a given frequency, rigorously termed insertion loss (IL)[1]

---

[1] **Remark:** *Attenuation vs. Insertion Loss.*

Strictly speaking, "attenuation" is the ratio of the EMI voltage (or current) at filter input to EMI voltage (or current) at its output. Insertion loss (IL) being the ratio of the voltages (or currents) at the load w/o filter and with the filter in place is a more exact definition for a filter performance. It eliminates the possible errors due to the various elements of the circuit: wiring, connectors, source and load impedances, and so on. However, since attenuation or rejection are most commonly used, we will keep "attenuation" along this article.

© The Author(s), under exclusive license to Springer Nature
Switzerland AG 2025
M. Mardiguian, *ElectroMagnetic Compatibility*,
https://doi.org/10.1007/978-3-032-02688-0_8

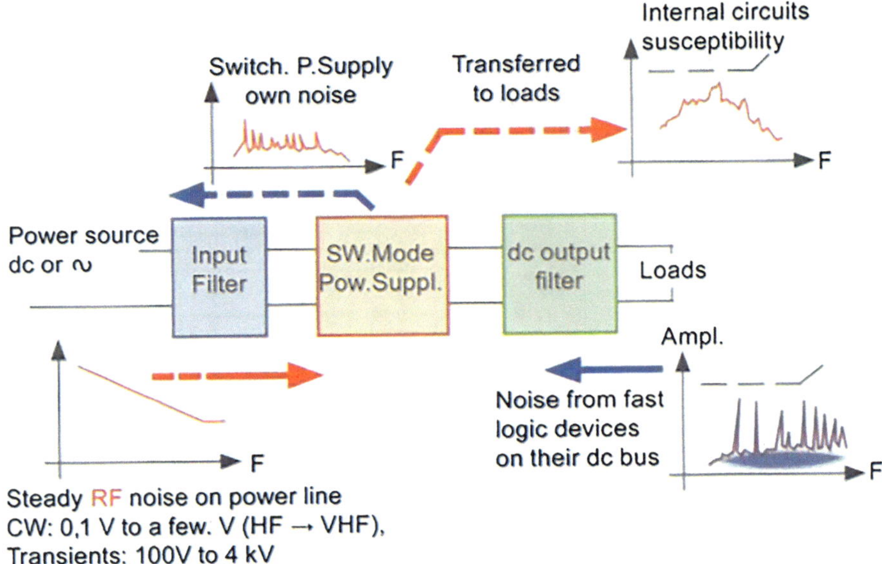

Fig. 8.1 The role of the input filter regarding power supplies immunity and emissions

(b) the cut-off frequency

(c) the number of poles, which itself depends on (a) and (b)

(d) the impedances on the source and load sides, that in turn will guide the filter scheme

(e) the type of filter: simple capacitor, simple inductor, L-C, Tee or Pi

(f) the normal service voltage/current, to be considered for the filter capacitors and inductors

8.2   Influence of Actual Source and Load Impedances

A same filter will exhibit different performances, depending on the values of Z_s, Z_L, and the impedances on the source and load sides. Because of this, one must beware of filter vendor's data. They are generally measured in a 50 Ω/50 Ω set-up (per Mil Std 222), and real in situ attenuation may differ significantly, to such extent that the performance shown on the data sheet is probably the only one you will never get! For answering the question: why a Capacitive, "L," "L, C," Pi or a T? There is a simple, flawless rule for choosing L, C elements, which suffers no exception:

With filters, capacitors should look toward high impedances, both sides, inductors should look toward low impedances, both sides.

A justified selection can be made with Fig. 8.2 showing the preferred choice for the four possible impedances' configurations. The borderline for LOW/HIGH impedance designation is 100 Ω.

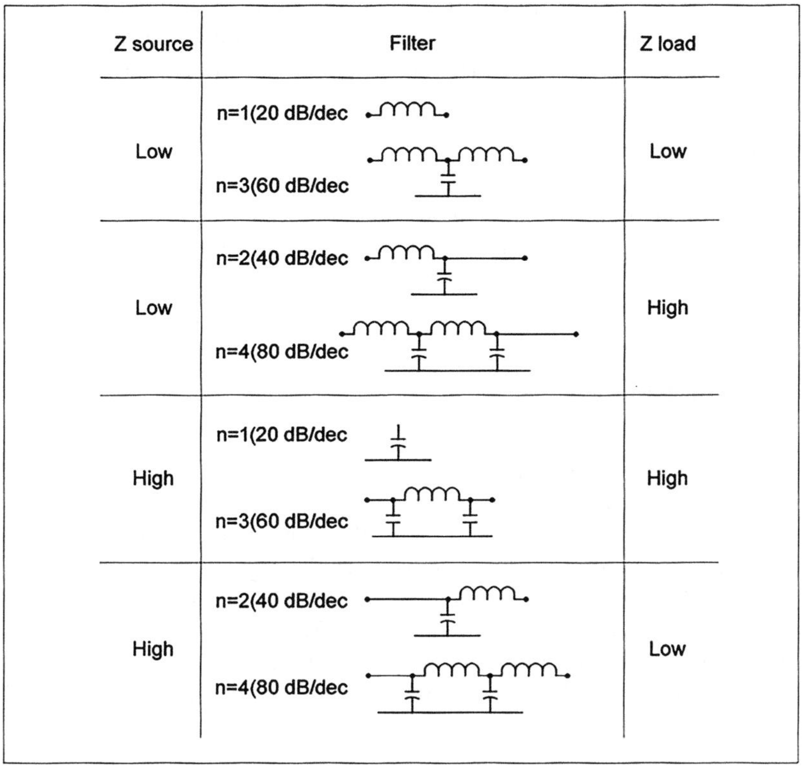

Z source	Filter	Z load
Low	n=1(20 dB/dec n=3(60 dB/dec	Low
Low	n=2(40 dB/dec n=4(80 dB/dec	High
High	n=1(20 dB/dec n=3(60 dB/dec	High
High	n=2(40 dB/dec n=4(80 dB/dec	Low

Fig. 8.2 Preferred filter arrangements according to input/output impedances. The High/Low border line is approximately 100 Ω

8.3 Cut-Off Frequency and Number of Elements

The number of poles in a filter (that is the number of L, C elements being cascaded) is determined by the desired attenuation at frequency F_{emi}, and by F_{co}, the cut-off frequency (or -3 dB point), below which the filter has no attenuation.

At a first glance, it would seem, for instance, that an ideal powerline filter should give no attenuation at all for 50/60 Hz, and start attenuating any undesired frequency above say 100 or 150 Hz. This would be a perfect, but enormous and expensive filter, due to the physical size of the capacitors and inductors. These are only found near large loads (greater than tens or hundreds of kW) for correcting the power factor of heavy inductive loads, or reducing harmonic distorsion on the utility side, upstream. Generally, the cut-off frequency of line filters for individual equipment has F_{co} values in the 10–30-kHz range. Figure 8.3 shows the attenuation of any filter, given its cut-off F_{co} and number of poles (n).

Numerical Example 8.1
From Example 7.2 in Chap. 7, we need 54 dB at 350 kHz. Assuming a cut-off at 10 kHz, how many poles our filter should have?

Answer: From Fig. 8.3, we see that, given F_{emi}/F_{co} = 350 kHz/10 kHz = 35, a first order (n = 1) filter is not enough. The filter must be at least a second order (n = 2).

8.4 Power Line Filters

Power line filters have been already partially addressed in former Chap. 7, "Power Mains Coupling." What is explained here is the optimization of line filter components for such specific application. Practically no modern equipment, with its fast digital circuits, switch-mode power supply regulators and eventually RF devices, could meet EMC requirements without an efficient power input filter. Switchers are the main cause of conducted EMI, violating the Military and Civilian RFI limits by as much as 40 or 50 dB for unfiltered items. Reciprocally, the power mains are carrying a multiplicity of EMI noise whose causes are out of our control that

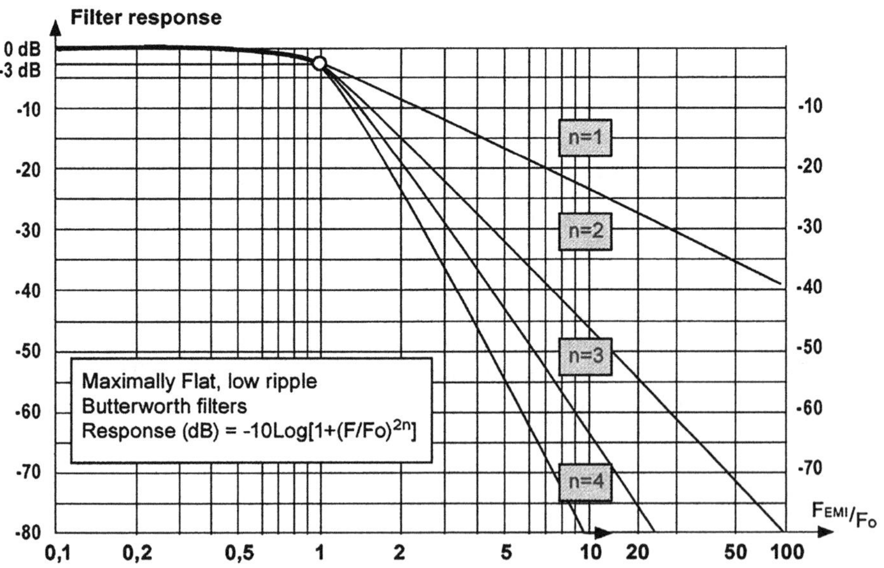

Fig. 8.3 Attenuation of low-pass filters normalized to the F_{emi}/F_{co} ratio for several values of "n"

equipment must tolerate without malfunctions or errors. In both cases, the power line filter is a key element for EMC compliance.

For EMI immunity, the power filter attenuation must be such as the most susceptible circuits inside the equipment do not receive on their power input a noise voltage greater than their threshold of sensitivity:

$$\text{Attenuation}: A(dB) = 20\log\left[\, V_{emi}\left(w/o\,\text{filter}\right)/V_{load}\left(\text{with filter}\right)\right]$$

This after having accounted eventually for the built-in attenuation of the power supply regulator.

Example 8.2
An equipment must not exhibit malfunction when submitted to bursts of 1 kV (CM) 300 kHz ringing transients on its 230 V AC input. The noise immunity of internal digital circuits is 0.5 V on their dc input. The existing filtering on the regulator output provide already 20 dB attenuation. What is the required filter attenuation?
Answer: A(dB) = 20 Log (1000/0.5) − 20 dB = 66 − 20 = 46 dB @ 300 kHz.

For Emission control, filter attenuation must be such as the HF noise generated by the equipment on its power input (CM and DM) does not violate conducted EMI emissions limits for AC or DC power mains:

$$\text{Attenuation}\left(\text{or Insertion Loss}\right): A\left(dB\right)$$
$$= V_{emi}\left(dB\mu V, w/o\,\text{filter}\right) - V\left(dB\mu V, \text{spec.limit}\right)$$

8.4.1 Optimizing a Filter Against EMI Emissions on Power Line

One earliest cause of user-created noise found in the power mains has been the ac–dc rectification and gate-controlled rectifiers used in many equipment. Increasing the use of high-frequency converters like switch-mode power supplies and inverters, variable speed drives, light dimmers, fluorescent lights, has shifted the noise emissions toward higher frequencies, causing both conducted and radiated EMI issues. In this respect, EMC filters not only can reduce conducted interference up to ≈30 MHz but they are also acting against radiated emissions from the power cord at frequencies above 30 MHz. Several mechanisms, internal to the switcher, are contributing to these emissions (Fig. 8.4).

- *Common Mode emissions caused by switch-mode power supplies*

  The CM conducted emission coupling is often the major cause of limits violation, for equipment supplied by 115 or 230 Vac. The CM leakage current is caused by the stray capacitance of the switching transistor, IGBT or MOSFET to their heatsink or nearby chassis. If the power supply is an isolated-type, a second

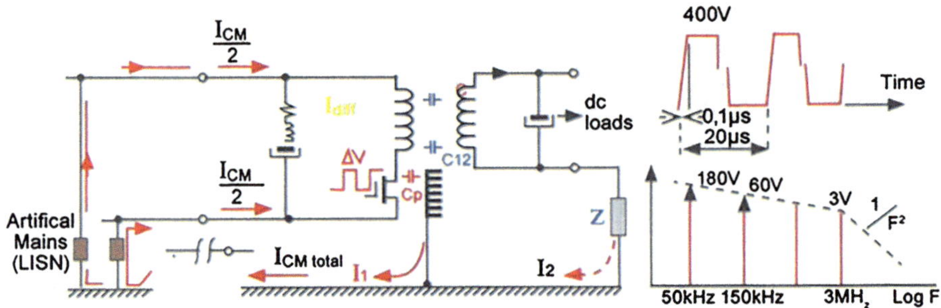

Fig. 8.4 Simplified view of CM emission path, with switched voltage waveform and Freq. spectrum for Example 8.3

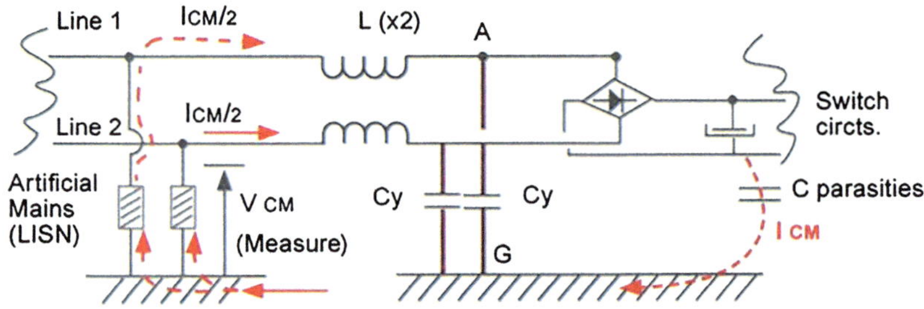

Fig. 8.5 Cut-away view of the front-end AC input, showing the filter Common Mode L, C elements

stray capacitance (C_2) may exist through the primary-to-secondary barrier of the transformer. Leakage current (I_{cm}) closes by the chassis ground, returning via the power mains impedance then back to the equipment Ph and Neutral input wires. For testing purposes, the noise level is measured at the artificial network (LISN) RF socket (Fig. 8.5).

Numerical Example 8.3

A 150 W switch-mode power supply operates at a frequency of 50 kHz. Influent parameters are:

- Stray capacitance (C_1) between the switching transistor and chassis: 100 pF.
- ESR of primary buffer capacitor on rectified 230 V input: 0.15 Ω up to 1 MHz.
- Primary current for full load, estimated: 1 A with 230 V ac, or 2 A with 115 V.
- Corresponding worst case peak switched current (115 V case): 3 A with 43% duty cycle.
- European EN 55032 class B limits:

  - 150 kHz: 56 dBµV
  - 1 MHz: 46 dBµV

Per safety regulations, the CM filter capacitors (AC line-to-ground) should not exceed 5 nF each, keeping the ground leakage current <0.5 mA.

Define the necessary filter for EN55032 compliance.

Answer: For the first harmonics of the switcher up to a few MHz, the CM curren is driven by the dV/dt of the switches into the stray capacitance C_1 and eventually C_{1-2}, behaving as a high impedance source (current source). Considering the sole CM filtering action, the simplified switching circuit shown in Fig. 8.4 results in the spectrum of CM current in the Ph/N wires (line #, I_{cm} on Table 8.1). Up to a few MHz, with the load being the two 50 Ω LISNs in parallel, the source impedance is dominated by that of the 100 pF stray capacitance C_1 (1600 Ω at 1 MHz). The Fourier plof the 400 V switched voltage gives a 50 kHz fundamental of 180 Vrms, with a third harmonic of 60 Vrms. Beyond 3 MHz, the spectral amplitude decreases sharply like $1/F^2$ thanks to the 100 ns rise time.

Bottom line on Table 8.1 provides worst case Class B limit violation of 44 and 58 dB at 150 kHz and 1 MHz, respectively. Interference at the fundamental 50 kHz frequency was ignored since there is no civilian regulatory limit below 150 kHz. Entering a default value $F_{co} = 10$ kHz for the power line filter, we see on Fig. 8.3 that for a F_{emi}/F_c ratio = 150/10 = 15,

- a single element (n = 1) filter cannot match the 44 dB need at 150 kHz,
- a two-element (n = 2) filter with L looking toward the LISN and C looking toward the high impedance (according to Fig. 8.2) will be all right (47 dB).

Thus, the n = 2 filter should have at least line-to-chassis (CM) capacitors C_y looking toward the high impedance side, with their value limited by the safety constraint to 2 × 5 nF maximum.

The value of the second element L is determined by looking at the simplified CM path (Fig. 8.5): with the two branches (I_{cm_1}) and (I_{cm_2}) being in parallel, the CM filter capacitors forms a divider bridge with the leakage capacitance C_1.

The corresponding voltage reduction factor is: 100 pF/(2 × 5000 pF) = 0.01.

Table 8.1 Calculation steps for CM filter attenuation of Example 8.3

Frequency	50kHz	150kHz	1MHz	30MHz
1) Harmonic Ampl. (Vrms)	180V	60V	10V	0,03V
2) Impedance of Cp		10kΩ	1,6kΩ	
3) 0,5 x Z $_{LISN}$		15Ω*	25Ω	
4) Icm		6mA	6mA	
5) V_{LISN} (mV)		90mV.	150mV	
(dBµV)		*100dBµV*	*104dBµV*	
6) Class B limit (dBµV)		56	46	
7) Limit violation (ΔdB)		**44dB**	**58dB**	

[a] **Actual LISN impedance (see Chap. 7) has been used, since its final 50 Ω value is only reached at 1 MHz**

The 150 kHz harmonic of the switch. Transistor voltage being 60 Vrms (line 1 on Table 8.1), the reduced voltage $V_{A\text{-}G}$ across C_y capacitors is: $V_{A\text{-}G} = 60$ V(rms) $\times 0.01 = 0.6$ V or 600 mV.

This reduced voltage is in turn applied to the network of filter inductance L_f in series with the two LISN in // represent an impedance of 30 Ω/2 at 150 kHz. Given that class B limit at 150 kHz is 56 dBµV or 0.6 mV, we find that the series impedance of filter inductance should be:

$L\omega = 2\pi F$. $L \geq 15$ $\Omega \times (600$ mV/0.6 mV), hence $L \geq 16$ mH
Taking a factor 2 (6dB) design margin: $L = 32$ mH

Note There was no need to recalculate the filter elements for 1 MHz: although limit violation is greater, we know that a two-stage filter will provide 32 dB more attenuation at 12 MHz than for 0.15 MHz.

- *Diff. Mode emissions caused by switch-mode power supplies*

  Combining with the CM current, a DM current circulates back and forth in the power input wires, due to the input dc storage capacitor. Since this capacitor has parasitic series resistance and self-inductance (ESR, ESL) the flow of the main switching current creates a differential voltage (line-to-line) across the primary input. By looking (Fig. 8.6) at the switched current spectrum, we can determine the VDM voltage across the input DC storage capacitor that in turn appears across the two LISNs. Notice that the set-up of the norm reads the voltage at the RF output of only *one* LISN.

  A second calculation table is made for the DM contribution, assuming 0.15 Ω ESR for the capacitor.

  Contrasting with CM emission, the DM voltage across the low impedance ESR, typically <1 Ω below a few MHz, behaves as a true voltage source. Therefore, using Fig. 8.2, the filter should have at least an inductor looking toward the primary capacitor and a line-to-line (DM) capacitor looking toward the two LISNs in series. .

  Table 8.2 provides a need of 34-dB attenuation at 150 kHz, significantly less than for the CM case. Yet, here again, no single n = 1 filter could achieve 34 dB at 150 kHz (Fig. 8.3). Since the CM mode filtering capacitors (5 nF) C_1, C_2 are useless

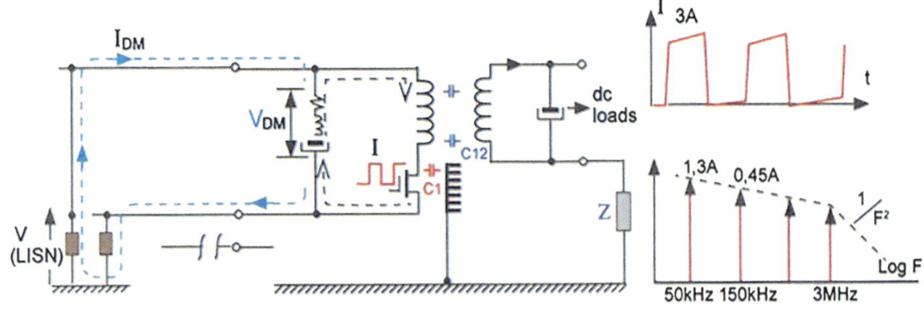

Fig. 8.6 Diff. mode path schematic with **switched current** waveform and frequency spectrum

Table 8.2 Calculation steps for DM filter attenuation

Frequency	50kHz	150kHz	1MHz
1) Current Harmonic calculated for a 43% duty cycle	1,3A(rms)	0,45A(rms)	0,06A (rms)
2) ESR of capacitor	0,15Ω	0,15Ω	0,15Ω
3) Vdm = 1) x 2)		68mV	10mV
4) 0,5 Vdm (at one LISN)		34mV	5mV
(dBµV)		90dBµV	74dBµV
5) Class B limit (dBµV)		56	46
6) Limit violation (dB)		34	24

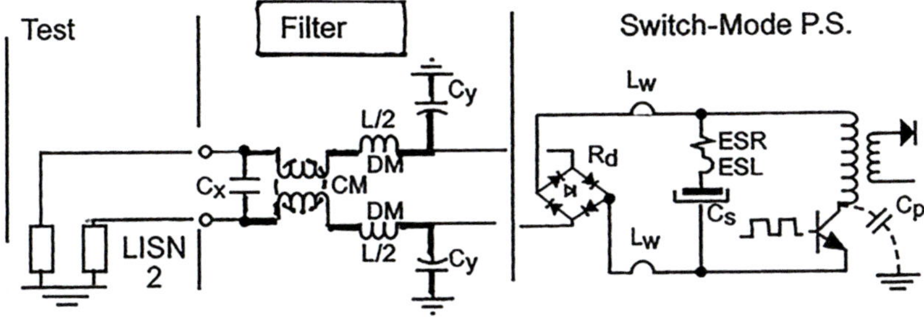

Fig. 8.7 Complete filter schematic, showing CM and DM elements

against this DM noise, a 300-µH inductance value, teamed with a 0.47- µF DM capacitor will provide 43-dB attenuation at 150 kHz, giving a 9 dB above the actual need. This DM inductance can be split into two windings, one for Phase and one for Neutral, avoiding a primary unbalance on the AC input. Or a single winding can be inserted on the dc high voltage side, after the rectifier bridge. In many cases, the double DM inductor can be provided at no cost, by using the leakage flux of the CM bifilar filter choke, typically 1%–3% of the nominal CM value.

A corresponding appropriate filter structure is shown in Fig. 8.7, optimized against CM + DM emission. The DM inductance, made by the leakage flux of the double-wound CM choke, brings a substantial space and cost savings. An other source of savings can be the use of X-2Y capacitor (Fig. 8.8).

8.4.2   Quick, Coarse Approximation for 50/60 Hz Filter Components

By default of more accurate calculations, a simple rule of thumb can be used for conservative, maximum values for the filter elements. The rationale is that, for ac power mains, we do not want the filter capacitor to derive uselessly too much 50- or 60-Hz current from the power mains. All the same, our filter inductance should not

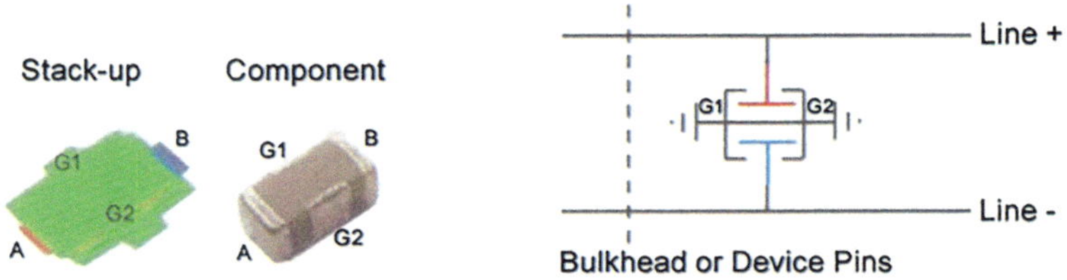

Very low Line-to-Grid parasitic inductance < 1nH : Grounding impedance < 1 Ω up to 300 MHz

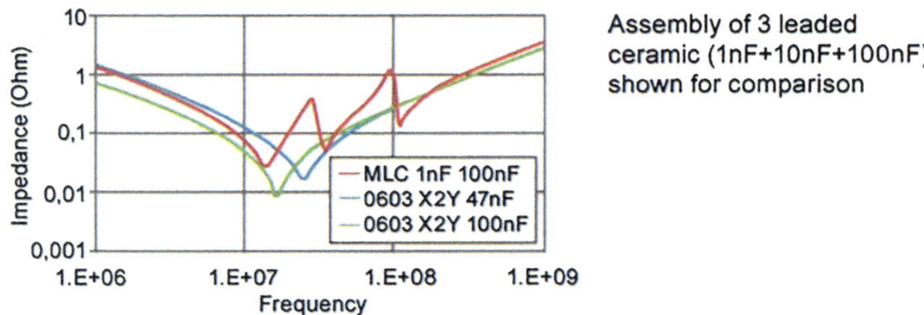

Fig. 8.8 Low-inductance "X2Y" capacitors (®) with combined CM + DM action. (Source Johanson-Dielectr)

cause unacceptable 50 or 60 Hz voltage drop. If we define 1% as a tolerable impact on ac current consumption and input voltage, we get:

$$X_c \geq 100\,Z_L \quad \text{and} \quad X_L \leq 0.01\,Z_L \tag{8.1}$$

X_c, X_L being the impedances of filter capacitor and inductance at the power mains frequency, and Z_L the equivalent load impedance. For a 50 Hz ac input, using more practical units for filter components:

$$C(\mu F) \leq 32 / Z_L \quad \text{and} \quad L(mH) \leq 0.032\,Z_L \tag{8.2}$$

If there are several capacitors like in a "Pi" filter, the formula applies to all the capacitors in parallel.

For the CM mode filter capacitors mounted line-to-chassis, which is to the earth connection, there is a safety constraint. In European countries, the maximum permitted 50-Hz leakage current is 0.5 mA for the most severe applications (domestic appliances), medical devices excepted. So, if used on 115 V/230 V equipment, these line-to-gnd capacitors, termed "Y" type are generally limited to 5 nF, with a 2.7 kVdc overvoltage survival test, since they can be exposed to mains voltage 100% of time.

Example 8.4

What are the maximum value of filtering capacitor (Line-to-Line) and inductor we can accept for a 230 V equipment with 10 A of normal line current?

Answer: The 10 A/50 Hz current can be translated into an equivalent filter load of 230 V/10 A = 23 Ω. Thus, maximum values are: C(μF) $\leq$ 32/23 = 1.4 μF, L(mH) $\leq$ 0.032 × 23 = 0.73 mH.

8.4.3   Filter Action Against Power Line Overvoltage Pulses and Filter Self-Resonance

EMC filters, being low-pass elements are unefficient against energetic pulses with duration exceeding a few microsecond. This is because long pulses have a large energy content in the low frequency range, where the filter has no attenuation. What is needed against energetic pulses is a component that is not frequency-selective but amplitude selective, like TVS or MOV. For more details, see Chap. 7 "Power line disturbances" and "Transient suppressors."

At the exact resonance of its inductor with its capacitor, a filter has no attenuation and may exhibit a gain instead, that is an overshoot (see Fig. 8.9 Curve A around 40 kHz) when the power is switched on or off. Typically, this overshoot manifests as a brief increase of the input voltage by 40%–100% of its peak value, with a possible stress or damage to the post-filter power supply components: rectifiers, electrolytic capacitors, and many more. This can be prevented by adding a low value damping resistor to the dc storage capacitors, or a transient voltage suppressor (TVS) after the filter, since this protection is often needed anyway for lightning and other AC mains transients.

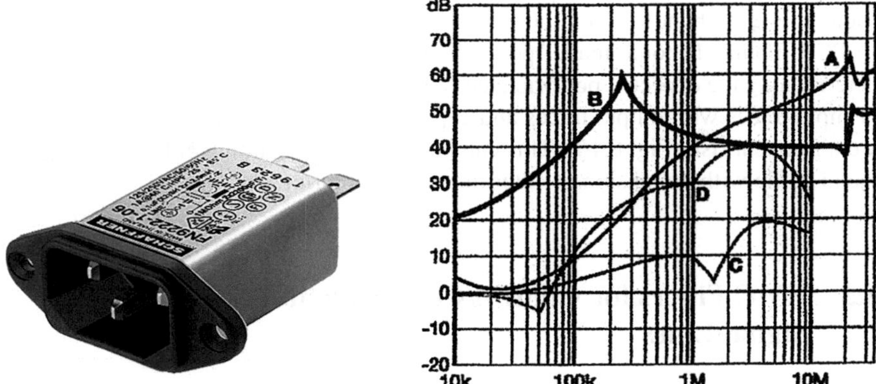

Fig. 8.9 Example of a commercial filter performance: DM (**A**) and CM (**B**). Besides the 50 Ω/50 Ω configuration, mismatch conditions are shown: (**C**) for 0.1 Ω/100 Ω and (**D**) for 100 Ω /0.1 Ω. The through-wall mount prevents parasitic coupling between the inner and outer sides of the equipment. (Source: Schaffner Co.)

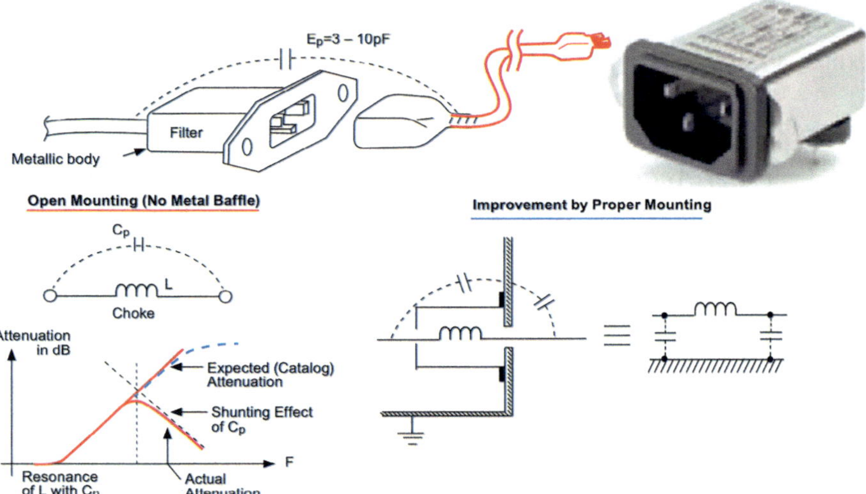

Fig. 8.10 Power line filter with through-wall mount. An open mounting (no baffle) is progressively shunting the filter choke by the stray capacitance. Here with Cp = 3 pF and L = 5 mH, attenuation deteriorates above 1.5 MHz

8.4.4 Recommendations for Filter Mounting

As much as its performances, the way a power line filter is mounted is crucial to its effectiveness. It should be installed as close as possible to the equipment wall opening where the power cord passes, the best being a through-wall mounting (Fig. 8.9). Metallic filter case is preferable, offering a tight metal-to-metal contact with the equipment box, or at least with a metal barrier. Wires on the line side of the filter should never be bundled with those on the load side.

PCB-mounted "home-made" power line filters are more prone to parasitic effects deteriorating the expected performance. Without a true metallic separation, crossing of input traces with output traces and capacitive coupling with other nearby traces are a frequent risk (Fig. 8.10). This can be prevented by putting at least a grounded metal baffle with dimensions > filter mounting footprint.

8.5 EMC Filters for Control/Command and Low-Level Signal Lines

Compared to power line filters, signal line filtering is a different challenge: signal filters must keep the desired signal unaffected in amplitude and phase, from dc up to the highest useful bandwidth necessary for signal integrity (Fig. 8.11). If not

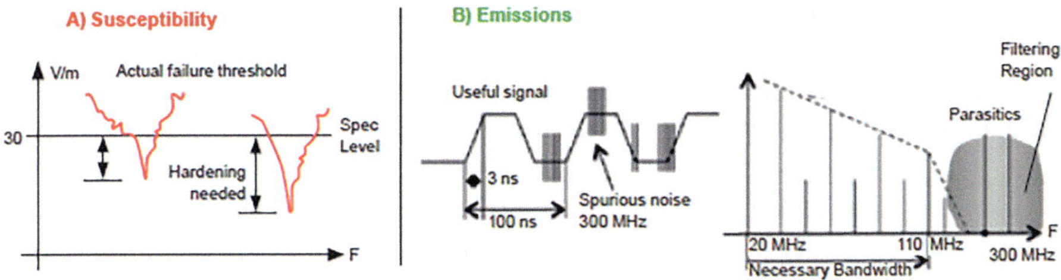

Fig. 8.11 Illustration of attenuation and cut-off frequency requirements for Susceptibility and Emissions

given by the datasheet of the signal technology, this useful bandwidth can be determined as follows:

Given the clock frequency or the transition time of digital pulse, the useful bandwidth, hence the EMC filter cut-off can be found as follows:

$$\text{if only the clock frequency is known}: F_{co} = 5 \times F(\text{clock}) \tag{8.3a}$$

$$\text{if only the transition time}\,(tr)\ \text{is known}: F_{co}(Hz) = 0.4/tr(s),$$
$$\text{or using more practical units}: F_{co}(MHz) = 400/tr(ns) \tag{8.3b}$$

These formulas for practitioners take into account a 20% margin for making sure that the useful signal suffers no detrimental distorsion in waveform and phase. From this, the filter can be entirely defined through the calculation steps described next. One must also define, from the actual interface (balanced or unbalanced), if the filter elements are arranged for simple DM attenuation or for CM attenuation.

Whatever the filter structure, the following conditions must be satisfied for an optimal filtering. Given the cut-off frequency F_{co}:

$$C = 1/\left(2\pi F_{co} \times (Z_s //Z_L)\right) \tag{8.4}$$

$$L = (Z_s + Z_L)/2\pi F_{co} \tag{8.5}$$

with Z_s, Z_L being source and load impedances

Example 8.5

Filter for reducing unwanted emissions from I/O signal (Fig. 8.11).

Fastest Digital signal to be filtered: 20 MHz Clock, tr = 4 ns, Zs = $Z_L \approx 100\ \Omega$.

It was found that a 300 MHz spurious noise riding over the digital pulse train must be attenuated by 24 dB. Define the proper filter.

Solution

F_{co}, from Eq. (8.3a) or (8.3b) = 100 MHz.

The filter should have no attenuation below 100 MHz, but must provide $\geq$24 dB at 300 MHz. Figure 8.3 shows that for a ratio F_{emi}/F_{co} = 300/100, neither a single pole (n=1) or two-pole filter are sufficient. It takes a n = 3 filter, that will produce a 28.5 dB attenuation. With 100 Ω being the transition for the "Low impedance/High impedance" criteria in Fig. 8.3, we can select either a "Pi" or a "T" structure. In any case, criteria of Eqs. (8.4) and (8.5) are used:

$$C = 1 / (2\pi F_{co} \times (50) = 33\,pF \quad L = (200)/2\pi F_{co} = 320\,nH$$

with a "Pi" structure, C will be split into 2 × 17 pF capacitors
with a "T" structure, L will be split into 2 × 160 nH inductors

8.5.1   One-Pole, Single L or C Filtering for Moderate Attenuation

When the needed attenuation is less than 30 dB, it is possible to get this figure with a single capacitor or inductor, bearing in mind that:

(a) this filter will be a one-pole type, with only 20 dB/decade attenuation above its cut-off frequency,
(b) it will work well only if Z_s and Z_L are in the same range of values: both >100 Ω for a capacitive filter, both <100 Ω for an inductive filter.

For instance, if one knows that $Z_L = 1000$ Ω, one common mistake is to assume that it is enough if the filter capacitor impedance Xc is <1000 Ω to get a filtering action. In fact, *no definition of the filter can be made without knowing at least the module of Zs and Z_L*. Cut-off frequency and attenuation of one-pole filters can be easily found in the following table (Table 8.3):

8.5.2   Ferrites

Inductive filtering by ferrite toroids, beads and sleeves deserve a special mention. While often regarded as the miracle, last-minute EMC remedies, they can be integrated in the design as part of the filtering line of defense. One interesting feature is that they are generally "lossy" ferrites, whereas some of the EMI energy is wasted

Table 8.3 Cut-off frequency and attenuation for one-stage (n = 1) filter

	F_{co}	Attenuation for $F > F_{co}$
Capacitive filter	$1/(2\pi Z_p C)$	20 Log $(2\pi F.C.Z_p)$
Inductive filter	$(Z_s + Z_L)/2\pi L$	20 Log $[2\pi F.L/(Z_s + Z_L)]$
	with $Z_p = Z_s // Z_L$	

in heat. They can be used as DM as well as CM blocking (two conductors in the same ferrite hole). Passing a wire "N" times in a ferrite hole will act as N^2 ferrites. The relevant ferrite parameter is its impedance/frequency curve (Fig. 8.12), allowing to predict its attenuation by:

$$A(dB) = 20\log\left[(Z_b + Z_s + Z_L)/(Z_s + Z_L)\right] \tag{8.6}$$

with Z_s, Z_L source and load impedances, Z_b: ferrite impedance

8.5.3   HF Limitations of Filter Components

Filters made of discrete components have their own parasitic problems. As frequency increases, inductors tend to behave as capacitors because of their internal winding capacitance and the stray capacitance between their leads. Capacitors tend to behave as inductors because of their terminal leads parasitic inductance. In a sense, above a certain frequency each L or C element start doing exactly the reverse of what is expected, spoiling the filter attenuation down to 0 dB. Careful filter components mounting may shift the problem to higher frequencies, but it always exist (Fig. 8.13). The only filter that never has parasitic response is the feed-thru capacitor, as seen next.

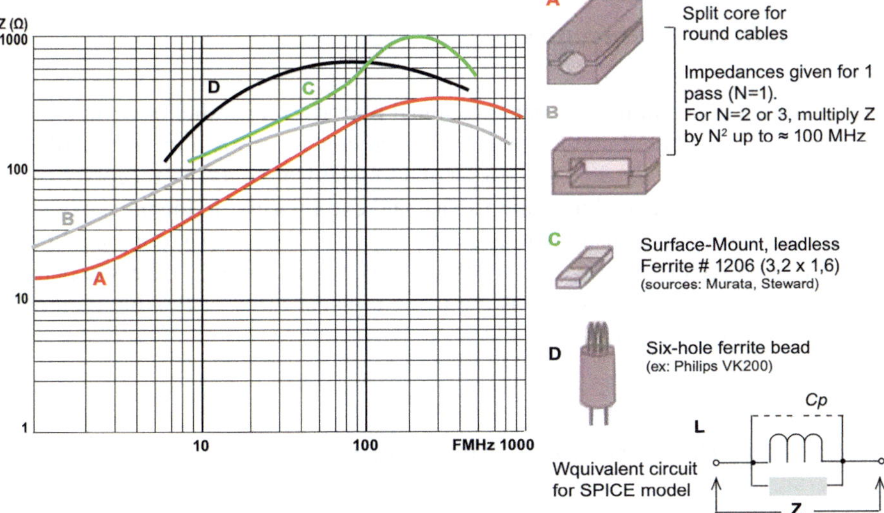

Fig. 8.12 Performances of various EMC ferrites

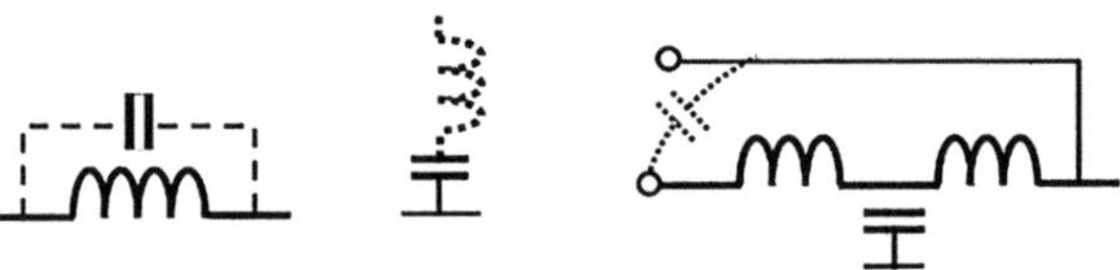

Fig. 8.13 A few typical causes of filter performances degradation by its own parasitic elements

8.5.4   Feed-Through Filters

Feed-through filters are based on a coaxial configuration of capacitor that suffers no parasitic inductance at all. The dielectric material is filled-in between the center pin electrode and the outer cylindrical electrode. This outer armature can be soldered, screwed or press-fitted into the chassis or PCB ground plane. Thanks to this design, the filtering is that of a perfect capacitor, with practically no frequency limitation. The coaxial capacitor can be teamed with ferrite beads to form a Pi or T filter (Fig. 8.14A).

Since their mounting requires some mechanical work, they do not lend to cheap, mass-production items. A low-cost version exists in the form of three-terminal, surface-mount capacitor known as "*semi-feed thru*." Although a no-match to the real feed-thru, it offers decent performance, provided the center terminal pad is soldered to the PCB ground plane by a direct via, or better two via in parallel for even less parasitic inductance (Fig. 8.14B).

8.5.5   Filtered Connectors

For filtering a large number of identical signal lines (like a parallel bus), an interesting option is the filtered connector, where each contact pin is a miniature filter. A single filtered connector can replace a set of a standard "n-contacts" receptacle + "n" discrete filters + associated mounting space and extra wiring. In the most expensive types, each contact is constructed as a tubular ceramic feed-thru, combined or not with a tubular ferrite to form a two or three-stage filter (Fig. 8.15). Less expensive, yet efficient "space-saver" versions are using a planar array or flexible membrane with embedded ceramic capacitors.

8.5.6   Mounting Precautions with Signal Filters

More than any others, filters for high-speed signals (>30-MHz bandwidth) require tight constraints for PCB mounting. Input-to-output traces crosstalk should be avoided and checked both horizontally (on a same layer) and vertically

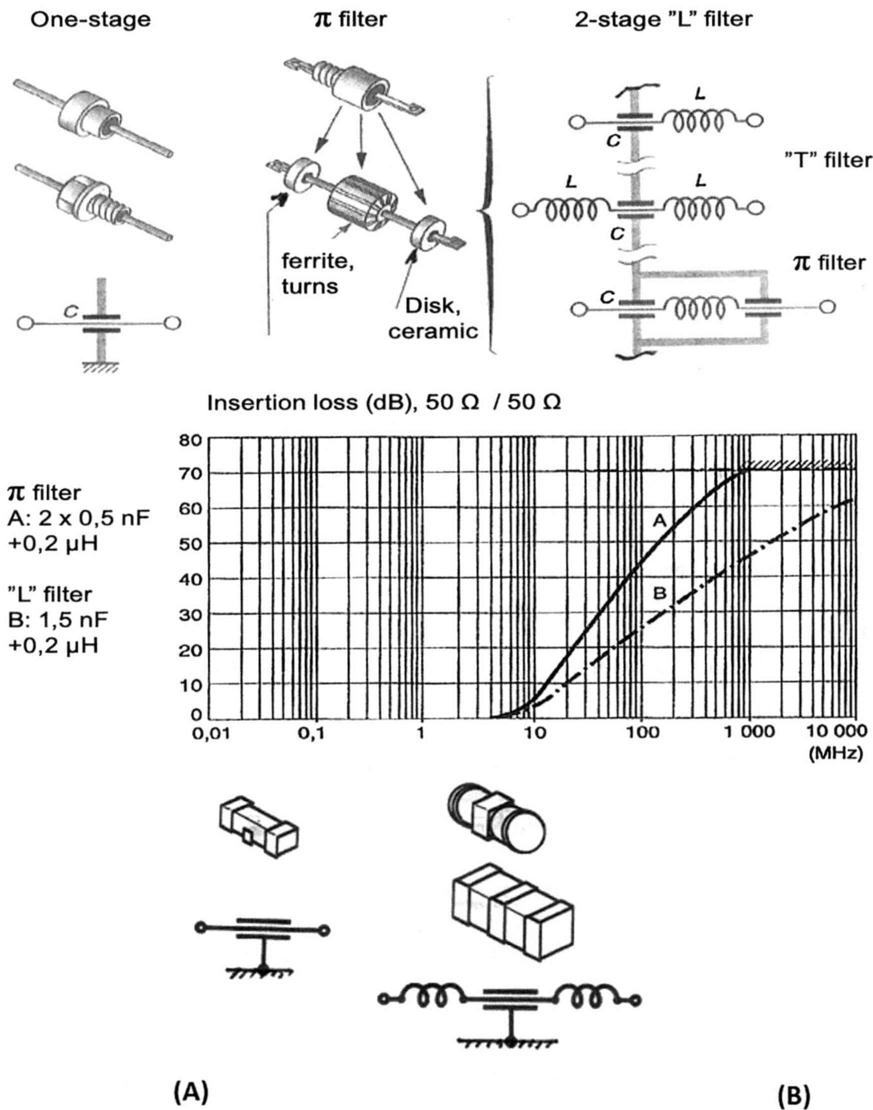

Fig. 8.14   Feed-thru filters: (A) classical type, (B) semi-feedthru

(layer-to-layer). When changing layer after leaving a filter, a filtered trace should not run close to an internal, unfiltered trace. This applies to the inter-layers via as well.

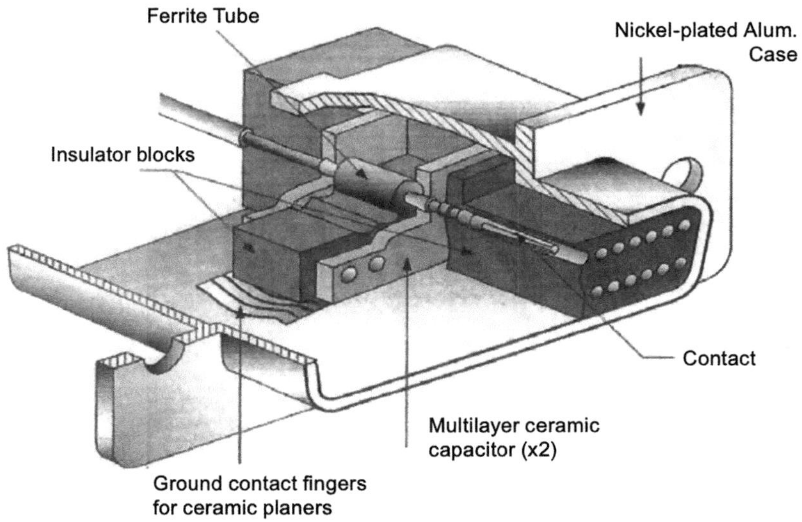

Fig. 8.15 Filtered connector

Quiz

There is ONE good answer, or ONE that is better than the others.

1. Cut-off frequency of EMC filters

    (a) an EMC filter should be selected such for a desired attenuation around its cut-off frequency
    (b) an EMC filter should be selected such as its cut-off frequency is the highest possible
    (c) an EMC filter should be selected such as its cut-off frequency is the lowest possible
    (d) the cut-off frequency depends on the source and load impedances

2. Capacitive filters

    (a) Capacitive filters are preferable when source and load are high impedances
    (b) Capacitive filters are preferable when source and load are low impedances
    (c) Capacitive filters are preferable when source impedance is low and load impedance is high
    (d) Capacitive filters must always have an inductance associated with the capacitor

3. The number "n" of elements in a filter

    (a) dictate its cut-off frequency
    (b) dictate the slope of its attenuation
    (c) dictate its characteristic impedance
    (d) should never be an even number

4. Signal filters for digital input/output lines

    (a) should be tuned to the clock frequency of the digital signal
    (b) should have a cut-off frequency below the clock frequency
    (c) should have a cut-off frequency greater than the signal bandwidth
    (d) should always be inserted on both + and − signal lines

5. Ferrites

    (a) ferrite beads affect only CM currents
    (b) ferrite beads on each + and − signal lines are affecting DM current only
    (c) one ferrite bead on an I/O pair will affect CM current with no effect on
         DM signal
    (d) N ferrite beads in series will attenuate RF current like N^2

Chapter 9
General Overview of the Civilian and Military EMC Norms

Preamble

Electro-Magnetic Compatibility is both a functional necessity—equipment must operate in their intended environment without being disturbed nor causing disturbances to other devices—and a society requirement: interfering with radio/TV broadcast, radio-comm., and other RF services is illegal, punishable by laws. Since unaware consumers cannot be easily prosecuted for using an equipment that is causing interference, an efficient way of controlling the situation is to apply stringent limits to manufacturers, such as the products put on market will "never" (exceptional situations accounted) cause interference. This latter aspect has been the driving force for civilian RFI regulations worldwide, since the 1950–1960 era. Similarly, non-interference is a critical concern for military systems, as well as civilian aeronautics and automobiles, because of the severe issues at stake, like mission failure and flight or driving safety if a radio-com. or navigation device is jammed.

All the same, it would be an unmanageable mess if too fragile (EMC-wise) equipment were put on the market, causing a monumental number of litigations between unaware consumers and vendors/manufacturers, or cancelations of contracts in the military trade.

Therefore, over the years, national and international bodies have devised complete series of emission and susceptibility EMC tests. The same was made in military domain by the various Depts of Defense. A fundamental difference is that Civilian Regulations are legally enforced, while compliance to Mil. Standards are a matter of contract fulfillment. In these Regulations and Standards, each one of the four facets of EMC domain (CE, RE, CS, RS) is covered by specific series of tests. The *Basic EMC Requirements* imposed *by law* or *Industry/Military standards* are:

- Electrical/Electronic Equipment/System must operate satisfactory in its intended environment.
- System must be self-compatible (intra-system EMC).

© The Author(s), under exclusive license to Springer Nature Switzerland AG 2025
M. Mardiguian, *ElectroMagnetic Compatibility*,
https://doi.org/10.1007/978-3-032-02688-0_9

- System must not interfere with neighbor systems.
- System must have a sufficient immunity to potential neighboring interference.

9.1   Susceptibility or Immunity?

Generally, EMC requirements call for a *no-error or no-damage* response under a threat of a given amplitude A1. On the other hand, the true *susceptibility can only be found* by increasing the level up to threshold (A2) where the equipment start showing malfunctions. The difference (A2 − A1) in dB is called the EMC margin, as the ratio between what the equipment *cannot tolerate* and the *demonstrated level of immunity*. A safe EMC margin is at least 6, and preferably 10 dB.

9.2   Military EMC Requirements

Military and Civilian EMC requirements are built on similar principles. But large differences exist in the severity of the limits and certain arrangements of the test set-ups. The apple core of Military EMC tests is the Mil-Standard 461. The fact that this Mil 461 Std was first released in 1967 and that more than 50 years later, it is still active, thanks to regular revisions made by competent Working Groups, is a sufficient proof of its efficiency.

At the time of this writing, the commonly applied Revision F has been replaced progressively by Rev.G (Jan 2016) which added specific lightning and ESD tests. It is one of the most widely used EMC tri-service standard, applied by Army, Navy, and Air Force in the majority of nations. NATO countries are often using STANAG 4370, a close equivalent to the US Mil-Std 461.The time-proven quality of its testing approach has convinced the civilian aircraft and the automobile industries to apply very similar EMC requirements.

Per the classification principles described at the beginning of this book, Mil-Std 461 follows the organization tree of Fig. 9.1. One of its many advantages is a foolproof check list of all the Emission and Susceptibility tests that are normally performed. Yet, all the tests shown in Fig. 9.1 are not necessarily carried-on, depending on the application of the equipment and the service that will use it: Army, Air Force, or Navy. The manufacturer must provide an EMC Control Plan describing how he will handle EMC during the design phases, and an EMC Test Plan showing how he will validate his design. Therefore an equipment data sheet declaring the item as "compliant to Mil-Std 461" has little meaning as long as it is not said which tests have been done and at which levels.

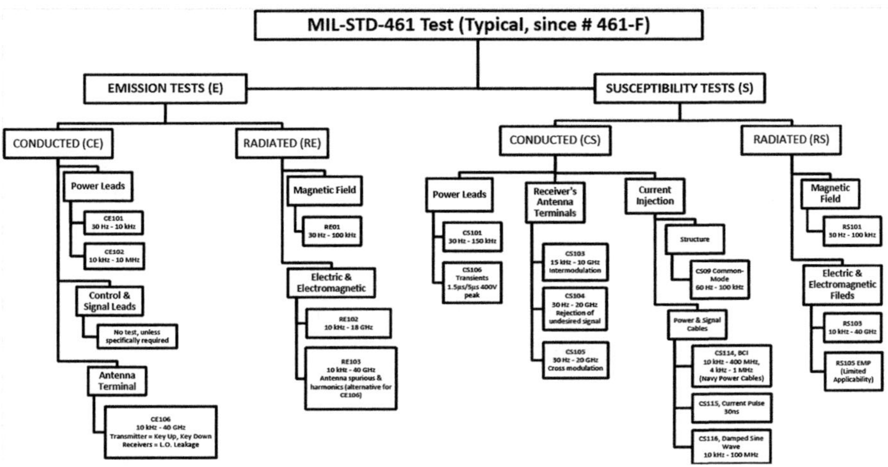

Fig. 9.1 Organization tree of Mil-Std 461 (Rev. E or F)

9.3   Emission and Susceptibility Tests Set-Up

An interesting feature of Mil-461 is that practically the same set-ups is kept for all the series of CE, CES, RE, RS tests, the changes being restricted to antenna types, receiver types and power generators for CW or Pulse excitation. Tests are carried in a shielded room (to avoid disturbances from or to the normal ambient environment). In addition, for avoiding parasitic reflections, the shielded room is made "anechoic" by covering walls and ceiling with RF absorbing materials like ferrite loaded pyramids or tiles, re-creating open field conditions. Since early 2000s, an alternative started to develop for testing susceptibility and emissions in a reverberating chamber. As of today, the anechoic chamber remains the most widely used.

The EUT is installed on a table with a copper foil top (Fig. 9.2), simulating a typical military mounting, in an aircraft, armored vehicle or ship, where cables are always laid not far from a conductive plane or structure. EUT cables are fixed on 5 cm insulating blocks, *which are forcing a fixed height above ground for all the tests to come.* In practice, the measuring instruments are located in an adjacent chamber, such as the operators and various peripherals used for the test control (PCs, printers, Video displays) do not interfere with the measurements.

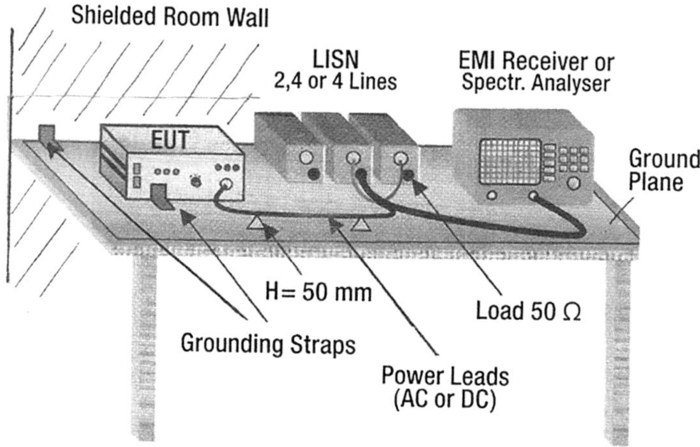

Fig. 9.2 Mil-Std set-up, geared for emission tests. The EMI Receiver is generally located in an adjacent anteroom

9.4   Mil-Std 461 Emission Tests

Mil-461 Emission tests ascertain that equipment installed on a platform, sometimes in large quantities, will not interfere with radio services on the host system (aircraft, vehicle, ship, etc.) nor with other systems sharing the same platform or military site.

9.4.1   Conducted Emissions Limits

The EUT must demonstrate that it does not generate on its external cables (power leads, but eventually signal cables in certain requirements) undesirable signals above the limits shown in Fig. 9.3. Limits are given in voltage (dBμV) picked-up at the measurement port (artificial network). In some specific cases, limit is given in current (dBμA), measured with an EMC current probe. Essential features of the set-up are shown in Fig. 9.2.

EUT accessories and peripherals that are part of its normal operation are also installed on the test table. The artificial network (see LISN, Chap. 7) is a very important device that simulates a typical impedance of the power mains, for both CM (Phase and N vs. ground) and DM (Phase-to-N) current paths. This prevent that a same EUT, tested in different places or labs could show different results because of varying impedances of the local power mains.

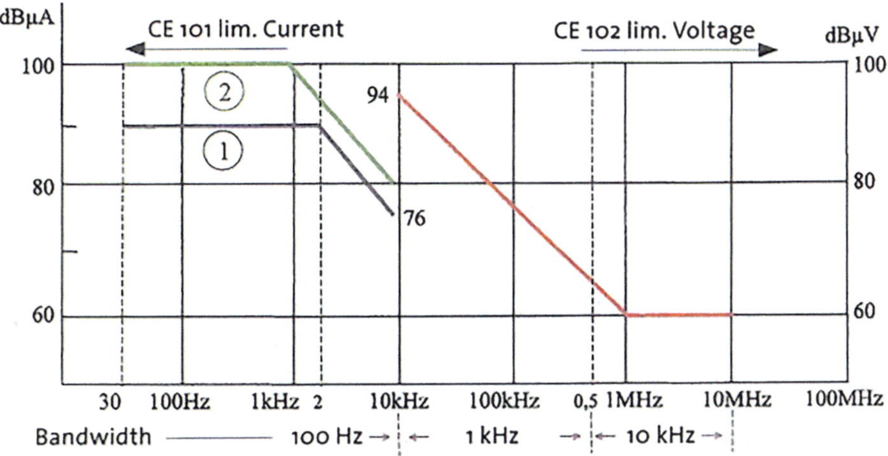

Fig. 9.3 Mil-Std.461-F, CE101/102 limits for the most severe categories (Aircrafts), AC or DC power leads. For unclear reasons, the limit stops at 10 MHz (former Rev. A, B, C had CE limits up to 50 MHz, a more wise frequency span)

9.4.2   Radiated Emissions Limits

The EUT must demonstrate that it does not radiate, by its box and external cables, undesirable RF fields above the limits of Fig. 9.4, showing the worst severity. The LISN is still there, such as the set-up is about the same as for the CE test, except that a calibrated receiving antenna is installed at 1 m distance (Fig. 9.5). The 1 m test distance (a difference with the civilian tests that will be described later) is justified by most typical military applications where various equipment are packed closely in the carrier. RE-102 test is a severe constraint, since it could require that at 1 m distance the EUT does not radiate more than 15 µV/m from 2 to 100 MHz, a rather tough goal to reach. If required, a low frequ. Magnetic Field emission (RE-101, 30 Hz to 100 kHz) is also carried with a loop antenna at close distance and various orientations from the EUT.

9.5   Mil-Std 461 Susceptibility Tests

Mil-461 Susceptibility tests ascertain that equipment installed on a given platform will operate properly in the most adverse conditions of its applications, in the vicinity of its host system radio/radar transmitters (aircraft, vehicle, ship, etc.) or eventually in hostile environment related to its mission. This includes also immunity to severe transient phenomena like lightning and NEMP.

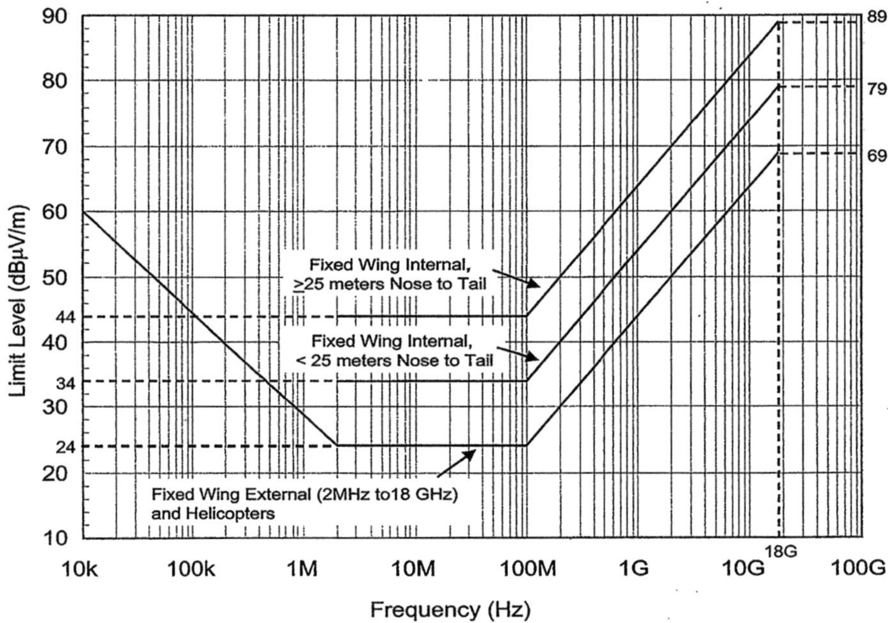

Fig. 9.4 Mil-Std 461 RE-102 limits for the most severe categories (Aircraft equipment)

9.5.1   Conducted Susceptibility Limits

The essential CS tests, in frequency domain (CW) or transient pulse (Time domain), are summarized in Fig. 9.6. All these tests are carried with the EUT mounted on the table bench ground plane, with power supplied through the LISNs, like for emission tests. For transient phenomena, the test plan must define what is an acceptable degree of EUT malfunction, if it is for instance a temporary, self-recoverable condition w/o critical consequences.

9.5.2   Radiated Susceptibility Limits

The principal RS tests are RS103 and RS105. *RS-103* is testing the EUT susceptibility to strong RF fields, as can be encountered in aircraft, military vehicle or ship environment. Besides the anechoic chamber, the test requires powerful amplifiers and antennas for creating RF field values of 20–200 V/m. Certain procurement specs. may demand a special immunity to High Intensity Radio Fields (HIRFs) as high as 2000–5000 V/m for pulsed radars (Table 9.1).

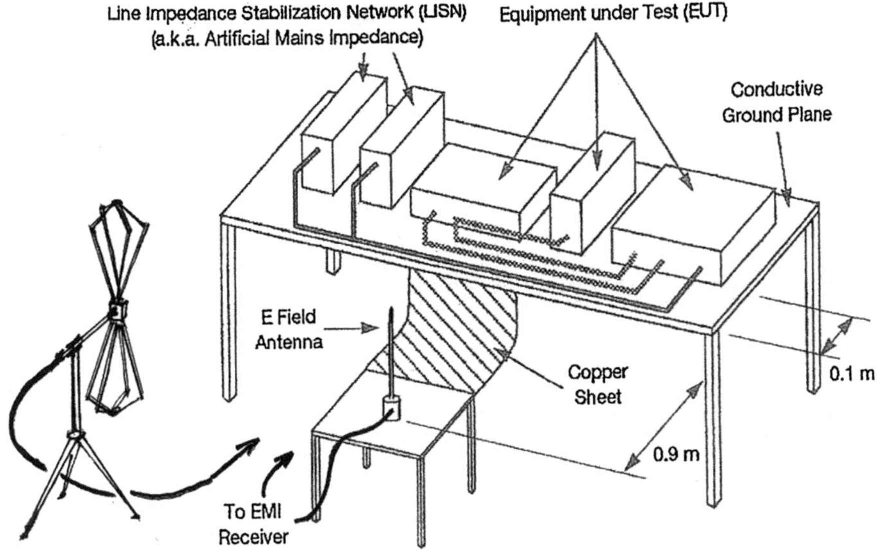

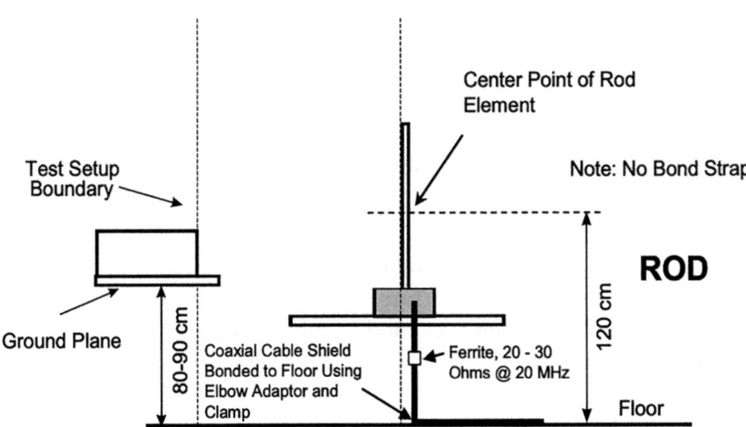

Fig. 9.5 RE 102 radiated test set-up. Receiving antennas can be vertical rod, biconical or log-periodic, eventually horn antennas, depending on the frequency bands. Vertical and horizontal polarization is required. Bottom, Mil-Std 461 F introduced a peculiar improvement to the 1 m rod set-up (10 kHz to 30 MHz): the dedicated 0.60 × 0.60 m ground plane—"the counterpoise" is no longer bound to the bench plane by a copper sheet, but virtually floating at HF, thanks to a ferrite rod on the coaxial cable to the receiver. This prevent uncertainties due to cable self-resonance with a stray capacitance to the ground floor

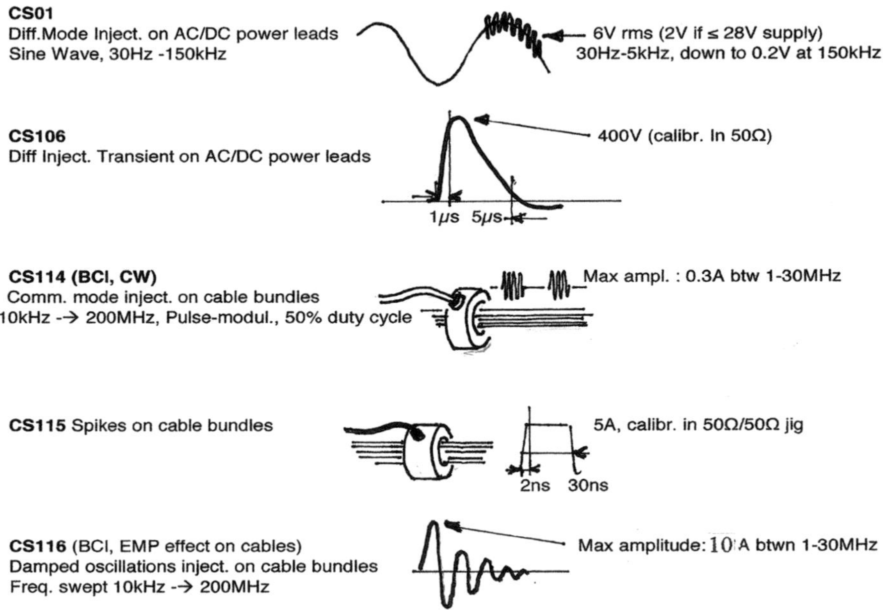

Fig. 9.6 Summary of Mil 461CS tests. The BCI tests are a substitute to an actual illumination of the entire system cables by very strong fields. The levels shown are generally those of the most severe category. CS106 (Pulses on power leads) has been deleted on Mil 461-G, replaced by CS118 pulse

Table 9.1 Normal RS-103 most severe levels for mission-critical equipment

Equipment in non-metallic aircrafts or Navy equipment above deck	Airborne equipment in protected areas or Navy equipment under deck
200 V/m, 10 kHz up to 40 GHz if justified	20–50 V/m, 10 kHz up to 40 GHz if justified

The *RS-105* is a very specific test simulating a strong radiated pulse, like a NEMP. The set-up is different from all the former ones, with the EUT exposed in a parallel plate or TEM chamber, which is excited by a high-voltage pulse to create a 50 kV/m field with 3 ns rise time and 30 ns duration.

9.6   Civilian EMC Requirements

EMC requirements for consumer and other civilian products are generally harmonized at International level, such as to avoid un-necessary multiplications of EMC tests when a product manufactured in country "A" is also exported for sales in countries X, Y, Z. These international standards are published under the International Electrotechnical Commission (IEC) authority, and studied/revised by committees like CISPR (mostly devoted to RF interference) or specific IEC working groups, as for ESD, Lightning, RF immunity, and so on. The compliance of an equipment to its relevant EMC requirements is attested by a mandatory marking, like CE for conformity to European Norms (EN) or FCC (for USA). We will only review here the major requirements that a non-specialist should at least be aware of.

9.6.1   Civilian Emission Standards

Civilian emission standards are a guarantee that equipment sold and installed for public use, sometimes in mass quantities, will not interfere with radio services. The most widely applied requirements are those of CISPR 14 (Household appliances) and CISPR 22 (data processing equipment), now replaced by CISPR 32, covering data processing and also radio, TV and video devices. Very similar limits are required by FCC part15-B "Low Power, unlicensed devices" in USA. These conducted and radiated limits apply to all equipment using functional frequencies >9 kHz, and are deemed sufficiently severe for radio protection in most critical conditions:

- close proximity of the RF source and victim receivers
- low attenuation of surrounding walls
- RFI sources/victims sharing the same power mains

More precisely, two classes were defined to control the above user's installation conditions:

Class B: a situation where source and victim can be as close (but no less) as 3 m, with only one separation wall in-between. They may share the same power distribution branch (Ph and Neutral) with only two power metering devices between them. That is the two equipment belong to different users, but one can be the next door's neighbor in a same building, typical of urban residential situation. This is regarded as a worst case for both Conduction and Radiation couplings.

Class A: a situation where source equipment is always located in industrial or large commercial sites, with no private dwellings at less than 30 m distance. The facility is powered by a dedicated step-down transformer that is not shared with private houses.

Accordingly, two corresponding severity levels are assigned for CE (dBµV) and RE (dBµV/m) limits. Class B limits are the most severe, since this configuration carries the highest risk of interference. As a result, an equipment that meets Class B limits can be used anywhere, w/o restrictions (Table 9.2). Conversely, Class A limits are 10 dB more permissive, but an equipment labeled "Class A" should not be used in residential areas. A note on the equipment case or the user's guide must warn clearly that:

"Usage in residential area could cause interferences, for which the user would be responsible".

Table 9.2 Compared severities of EMC regulations depending on the type of environment

EUT Environment	Emission Regulations	Immunity Regulations
RESIDENTIAL	MOST SEVERE	LESS SEVERE
INDUSTRY	LESS SEVERE	MOST SEVERE

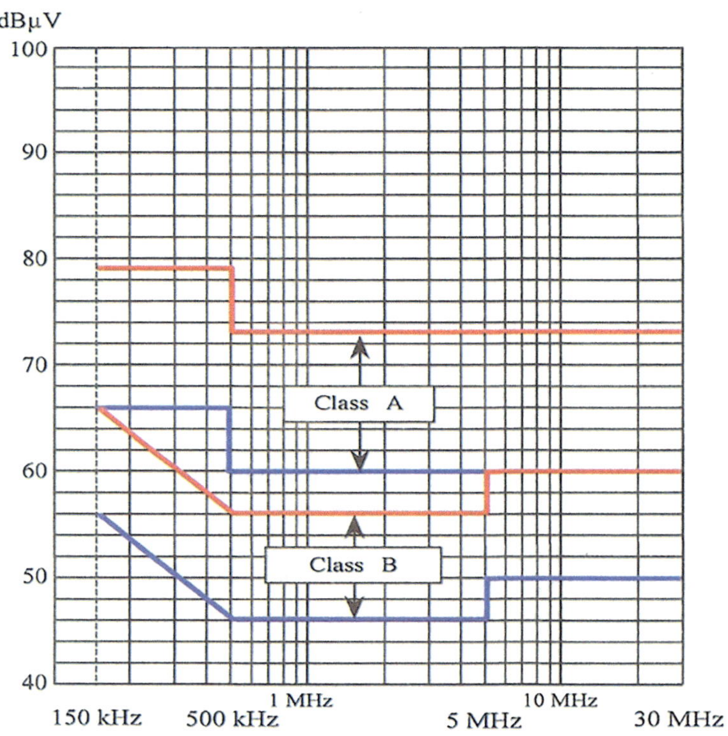

Fig. 9.7 Limits for EN 55022 Conducted Emissions. The red line captioned QP (quasi-Peak) is a relaxed limit authorized for those EUT emissions that are BroadBand (spurious noise with low repetition rate)

9.6.2   Conducted Emissions (Civilian)

The EUT must demonstrate that it does not generate on its external cables (essentially power leads, but also signal cables in certain applications) undesirable signals above the limits of Fig. 9.7. In contrast to Mil-Std 461 CE test, the limit is imposed up to 30 MHz, a safe practice. Essential features of the set-up are shown in Fig. 9.8 The EUT is installed in a manner representative of its normal use (table or floor-standing), with eventually its accessories and peripherals. Like for Mil-Std, an artificial network (LISN) is inserted on the EUT power leads for both CM (Phase + N vs. ground) and DM (Phase-to-N) current paths.

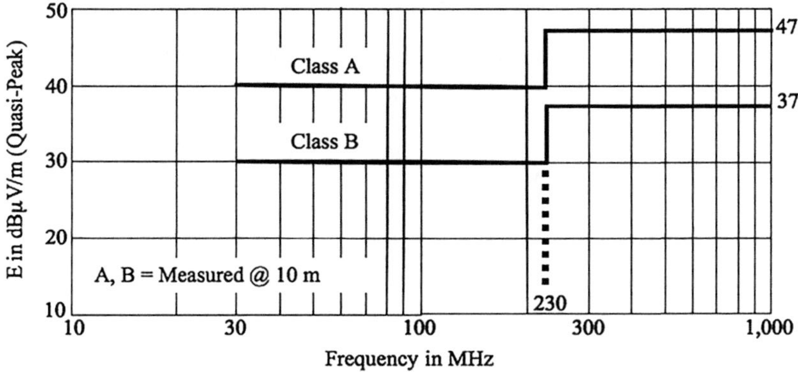

Fig. 9.8 Limits for EN 55032 (ex 55022) radiated emissions

9.6.3   *Radiated Emissions (Civilian)*

The EUT must demonstrate that it does not radiate, by its box and its external cables, undesirable RF fields above the limits shown in Fig. 9.8. The essential features of the test set-up are shown in Fig. 9.9. Like for the former test, the EUT is installed in manner representative of its normal use (table or floor-standing). The set-up is about the same as for the conducted test, including the LISN. While in theory it is supposed to be done on a free-field open test site, the test is generally performed in a shielded anechoic room. The calibrated antenna is either at a fixed height (for certain tests), but more generally on a mast that allows a variable height scanning to find the maximum emissions. For a same reason, the EUT is on a rotating plate, allowing a 360° scan for searching the maximum radiation pattern. These two refinements are causing more test time.

Both Class A and B limits are officially given at 10 m distance. For practical reasons, they are often measured at 3 m, for which a +10 dB adjustment is made to the "B" limit. For those EUT with clock frequencies >108 MHz, limits are extended as follows:

3 GHz, Class A: 56 dBµV/m/Class B: 50 dBµV/m.
3–6 GHz, Class A: 60 dBµV/m/Class B: 54 dBµV/m

Remark About CISPR/FCC Radiated Emissions Limits Most civilian requirements have RE limits starting at 30 MHz only. The rationale for not measuring radiated emissions <30 MHz is twofold:

(a) below 30 MHz (corresponding wavelength is $\lambda \geq 10$ m), the radiation from the EUT box itself is minimal due to the small size of its internal elements—typically less than 1 m. The remaining contributors are the I/O cables, but their length, too, is limited ($\approx$1.50 m) by the set-up, so they will not radiate efficiently in such set-up.

Fig. 9.9 Test set-up for **EN 55032** radiated emissions. (Courtesy DLS Systems)

(b) Measuring radiated emissions at 3 or 10 m in a limited test space for frequen-
    cies <30 MHz complicates the test, because of a mediocre efficiency of avail-
    able antennas, the near-field radiation conditions of the test, etc.

Both (a) and (b) difficulties are solved by the conducted emission measurement:
up to 30 MHz, the conducted limit is assumed severe enough to grant that the equipment will not interfere with nearby victims, not only through conduction on power
mains, but also because the corresponding cable currents will not radiate an objec-
tionable field.

9.6.4   Civilian Immunity Standards

The immunity (sometimes quoted as "susceptibility") standards are a guarantee that
equipment sold to consumers or industry will not suffer from unexpected malfunctions or failures from their intended electro-magnetic environment. These could
result in direct or indirect prejudices for the users, eventually with harmful consequences, not to mention a maze of litigations to decide who's fault it was.

The conducted and radiated immunity tests of the most common civilian standards (IEC 61.000-4 series) are summarized in Fig. 9.10. The EUT must demonstrate that it does not exhibit failures or malfunctions when exposed to these tests
levels. The severity levels, required by the European Directive and corresponding
European Norms (ENs) of the CENELEC are depending on the class of equipment
and its application, varying from ordinary consumer appliance up to industrial or
medical equipment, where failures and malfunction could have disastrous
consequences.

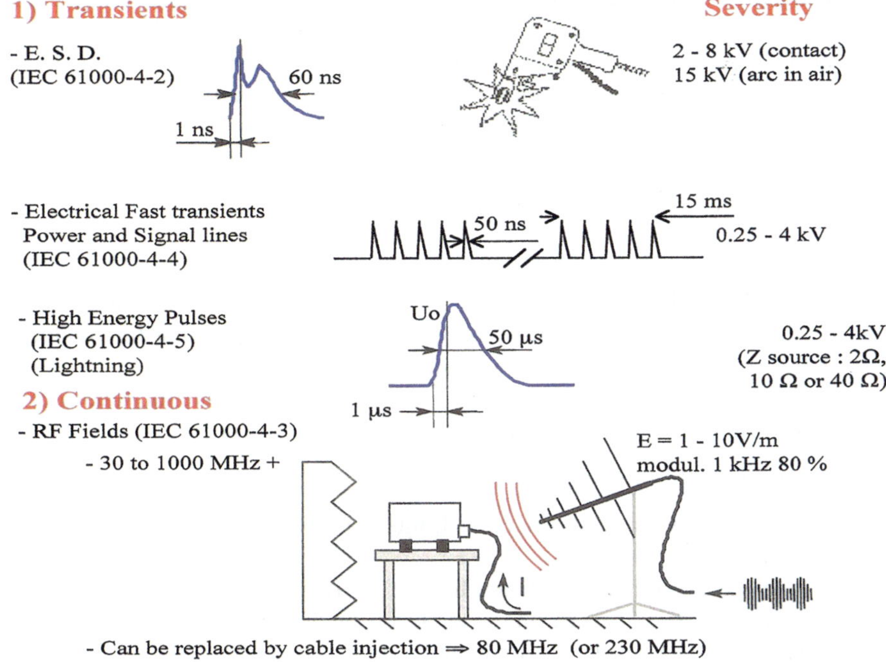

Fig. 9.10 Summary of Civilian Conducted and Radiated Immunity tests

9.7 Concluding Remarks About Military and Civilian EMC Requirements

A few interesting comments can be made when comparing civilian vs. Mil-Std EMC limits. While deemed more severe than civilian limits:

– *for CE tests*, the Mil-Std limits are practically the same as those of the civilian class A (less severe),
– *for RE tests*, the Mil-Std limits: could appear much more severe (limit about 20 dB more stringent than CISPR 22 class B, and measured at only 1 m). But the fact that the cables are laid 5 cm above a gnd plane is almost offsetting this difference up to ≈ a few hundred MHz. Because of this, a commercial off-the-shelf item that passes civilian Class B with a fair margin (≥6 dB) may be expected to meet the most severe Mil.461 RE-102. Of course, this does not exonerate of an actual testing.
– *for susceptibility*, the Mil-Std tests are generally more severe than the civilian ones. However some IEC tests like Fast Transient Bursts (61000-4-4) or ESD (61-000-4-2) that are very severe in the industrial category have no real equivalent in the Mil-Std 461 arsenal.

Having reviewed the essentials of Civilian and Military testing, readers may wonder *"Now, what to do if we fail the test?"* This will be the subject of the next chapter. This is a necessary stop-over for understanding what went wrong and what to do to attain compliance including fix/fretrofit an existing product which is experiencing EMC problems in the field.

Quiz

There is ONE good answer or ONE that is better than the others.

1. Mil-Std461 vs. Civilian EMC tests

    (a) Both Mil 461 and Civilian tests are using the same procedures and test set-up
    (b) Mil-Std461 vs. Civilian EMC tests practices are very similar for conducted EMI but differ significantly for radiated measurements
    (c) Civilian radiated tests require exclusively an open field site for radiated EMI
    (d) Mil-Std461 vs. Civilian EMC tests are totally different

2. Comparing EMI Emission tests

    (a) the most severe class Radiated limits for Mil. and Civil. are equivalent within 6 dB
    (b) the most severe class Radiated limits for Mil. and Civil. are equivalent but the civilian test is measuring worst levels by 360° rotation of the EUT
    (c) Mil-Std461RE levels are significantly more severe
    (d) No comparison can be made because of different test distances and cable heights

3. Comparing EMI Susceptibility tests

    (a) Military RS tests are made exclusively in an anechoic chamber
    (b) RS tests can be made in an anechoic chamber or reverberating chamber
    (c) Military RS tests can be replaced by a Bulk Cable Injection (BCI) test
    (d) Civilian RS tests are only required for harsh industrial or medical applications

Chapter 10
Troubleshooting EMI Simple Hints for Identifying and Fixing EMI Troubles

Preamble

This chapter covers the essentials of a domain that is seldom addressed in the current EMC literature: *"What to do when an equipment—or a whole system—is failing the tests or is experiencing Interference (EMI) problems"*?

Whatever we are dealing with a prototype that is failing on one or more EMC tests at the end of its development phase, or an already installed equipment that exhibit on-site problems, such situation must be solved quickly, with a product whose hardware cannot be deeply modified.

Contrasting with a development phase where many EMC solutions are available on a product that is still flexible, the engineer confronted to a failing equipment has to detect, diagnose and fix a problem that could be unpredictable, with elusive symptoms, troublesome and penalizing for the user. RFI, ESD, Transient surges, Crosstalk are complex threats involving many interactions. No human brain can see at a glance all the possibilities and limitations of the available solutions, where options are limited anyway.

Here we will help identifying an EMI problem and its coupling paths in order to correct it with fixes that must be quick, using components that are readily available and field-applicable, if necessary. All this using instruments and accessories that are portable, rugged, and relatively inexpensive, not requiring the sanitized environment of an anechoic shielded room.

.

10.1 Various Aspects of an EMI Problem

Assuming it was unexpected—in fact many times it WAS expectable, coming out of a design that neglected EMC, or resulting of some deliberate cost savings or lack of installation precautions—an EMI problem may show-up with different situations:

© The Author(s), under exclusive license to Springer Nature
Switzerland AG 2025
M. Mardiguian, *ElectroMagnetic Compatibility*,
https://doi.org/10.1007/978-3-032-02688-0_10

(A) The status of the equipment:

- equipment is a well-advanced prototype, or early pre-production item, or
- equipment is already commercialized, with little possibilities, if at all, for modifications

(B) The nature of the problem

- equipment is failing on one or several EMC mandatory tests, emissions or immunity
- the stand-alone equipment did not fail (or not yet) the EMC tests but creates functional problems when integrated in a system configuration
- the equipment is malfunctioning on-site, in certain installations only

(C) The occurrence of the problem

- problem is continuous or quasi-continuous (occurring frequently, in a repeatable manner) or
- problem occurs rarely, in a random, unpredictable manner

Each one of these A, B, C conditions, and eventually their combination will require a different approach, according to the urgency, cost and possibility of investigations.

Note We intentionally ruled-out the case of an equipment that is disturbing itself (Internal EMI), since such problem is normally discovered soon enough during development phase. Yet, this case can be analyzed using the routines described hereafter.

10.2   A Few Facts Leading to Troubleshooting Optimization and Time Saving

For both Emission and Susceptibility specifications, Conducted and Radiated aspects are treated separately, since the former are generally the dominant mode below 10–30 MHz region, while the radiated concerns are generally driving issues above 30 MHz.

10.2.1   Advantages of Early EMC Testing During the Design Phase

Statistics from EMC test labs reveal that 50% of the products submitted for final compliance fail the first time, at least on one test. Using the simple workbench tests described here, that statistic can be reduced to only 10% or 15% (H.Ott, [13]).

Although not as accurate as certified lab measurements, workbench EMC measurements are simple, inexpensive and can be performed early in the development phase, giving a preview of its EMC performance or weaknesses. They can be run in the designer's laboratory, with limited, relatively inexpensive equipment.

From now-on, the equipment of concern will be designated as Equipment Under Test (**EUT**).

When planning an EMI problem investigation, one should consider that:

(a) EMISSION MEASUREMENTS are faster, easier to do than susceptibility ones:

- You do not try making the equipment fail, you just let it run.
- When limit is exceeded, it is rather easy to trace the culprit source.
- No risk of damaging the EUT or associated equipment by excessive stress.
- Improvements you will apply are generally beneficial to immunity as well.

(b) CONDUCTED MEASUREMENTS are faster, easier to do than radiated ones: Set-up is simpler, requiring less instrumentation, less prone to measurements uncertainties or errors.

10.2.2   Before Starting Any Investigation

Make yourself familiar with the EUT features relevant to EMC: main frequencies of the digital signals and switchers, type of I/O interfaces (balanced or not), etc. Get a figure of how many dB of improvement are needed, at which frequency (or frequency range)? Having to harden a device by 6 or 60 dB (see Table 10.1) will put you on two different ball parks!

Table 10.1 Broad reminder of the essential amplitude and power ratios, and their dB equivalents

Amplitude ratio	Power ratio	Corresponding dB
(for current, voltage or field)		
2	4	+6
3.15	10	+10
10	100	+20
100	10^4	+40
1000	10^6	+60
0.5	0.25	−6
0.31	0.1	−10
0.1	0.01	−20

10.3   Part I: Troubleshooting Emission Problems

We are focusing here on **EMI Emission**, while further sections, Part II will cover **Susceptibility problems**, including those occurring in the field, where we miss the commodities and elbow room enjoyed in a development lab. According to our previous list, this choice is the faster to perform if you have such chance. Several situations may occur:

10.3.1   Prerequisites for Troubleshooting a Prototype or a Pre-qualification Item

Here, the EUT is designed, but some aspects are not completely frozen, thus room exists for minor changes. You are probably not (or no longer) on an EMC test site, and in any case an EMC test chamber is not the place for cut-and-try investigations. Yet, you will need a location with the following minimum characteristics:

- *A quiet RF ambient,* far enough (at least >3 m) from powerful noise sources like fluorescent tubes, air-conditioning compressors, elevators, power converters, etc. Ground level or basements, away from the building façade are preferable to upper floor locations.
- *Noise-free power mains,* with the EUT and associated test set-up being all fed from a same ac branch that does not supply other noisy equipment. Since this is often difficult to ascertain, a good precaution is to install an Isolation Transformer (IT) and an EMI filter near the distribution panel. In addition to a good isolation from the rest of the ac distribution noise, the IT also avoids tripping the Ground Fault sensor when you will connect a LISN (artificial network) for some tests.
- *A test ground plane,* extending at least up to and beyond the EUT footprint, associated cables and measuring instrumentation. This will be the artificial RF reference for the entire set (LISNs, Spectrum analyzer). Any solid metal sheet, not necessarily copper, aluminum or galvanized steel will do; thickness is not important. By default, heavy-duty kitchen aluminium foil can do, fold in double layer for tear-off resistance. It will also allow for a well-defined height-to-ground distance of the EUT cables, improving the test repeatability. For safety, the plane should be connected to the nearest accessible earth reference (earth bus of the room power panel for inst.)

All the instruments/accessories will be grounded to this test plane using wide, short straps. The EUT is simply grounded via its power cord safety conductor, if any. Do not ground the EUT chassis directly to the test plane, unless it is a normal practice for its use (f. inst. Military equipment).

10.3.2   Minimum Instrumentation for Checking Conducted and Radiated Emissions

Conducted Emission Specs generally cover the 0.15–30 MHz frequency range, with some military, vehicle or aeronautic equipment requiring 10 kHz up to 100 MHz coverage.

Radiated Emission Specs generally cover the 30–1000 MHz frequency range, for civilian regulations, extending eventually up to 6 GHz. Military, vehicle or aeronautic equipment may require 10 kHz up to 18 GHz.

Yet, as of today, it is very unusual to find an EUT exceeding *radiated emissions* limits below a few Megahertz or above a few gigahertz, except for rare cases of interference to sensitive UHF receivers. Given the above coverage the following test gear is a minimum:

- *Spectrum Analyzer*: your most useful, and expensive piece of equipment. Yet rugged, portable and easy-to use Spectr. Analyzers are available today for less than 2000€ (Fig. 10.1). As a minimum, the following features are needed:

  - frequency coverage of at least 0.1–1.500 MHz, or 10 kHz to 3 GHz if you have to check EUTs for military or airborne applications.
  - at least 10 kHz and 100 kHz selectable resolution Bwidths, with 1 MHz being also recommended.

Choose a model with *built-in tracking generator*, that will be precious for quick evaluation of some fixes (ferrites, filters, etc.) whose characteristics are unknown or doubtful. Prefer a model with N or SMA style RF input. BNC inputs tend to become undependable and leaky fort repeated use above 30 MHz. An internal LNA (low noise pre-Amp) is also a precious feature.

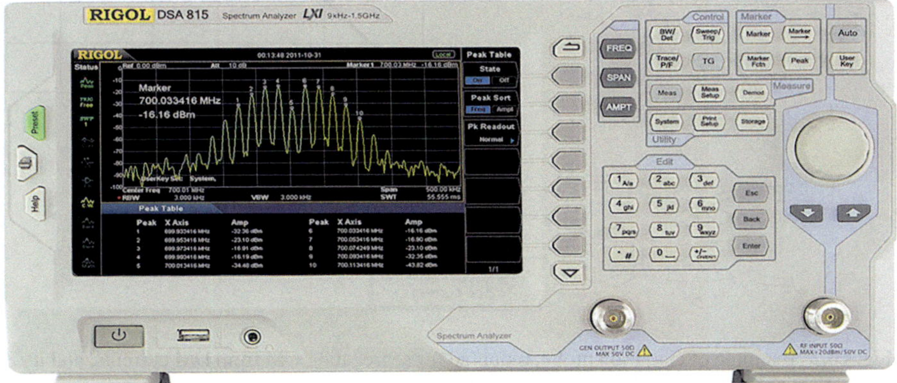

Fig. 10.1 Modern spectrum Analyzer, with N-style connectors for RF input, Tracking Gen., and handy, "intuitive" manual controls. Frequency scale can be linear or log. (Courtesy RIGOL Corp)

- ***EMI current probe***: a most useful piece, quite inexpensive. Select a model with well calibrated Transfer Impedance (Zt), preferably flat between 1 and 100 MHz and characterized up to 300 MHz (Fig. 10.2). Eventually one can make his own current probe from a snap-on ferrite ring [12], with a shielded core preventing pick-up of external fields.
- ***Low noise pre-amplifier***, with $\geq$20 dB gain and noise figure <4 dB. Although not necessary for powerline conducted emissions, it can be useful for detecting low EMI signatures <10 μV, especially against stringent Radiated Emission limits. Some Sp. Analyzers have one built-in.
- ***LISN*** (Line Impedance Stabilization Network). This important device simulates a standard, typical impedance of the power mains, for both CM (L_1, L_2 vs. ground) and DM (L_1 vs. L_2) current paths. It prevents that a same EUT, tested in different labs could show different results because of varying impedances of sites power mains distribution.
- ***Small proximity Magnetic field probes***. Field measurements are normally made with calibrated EMC antennas which are large (a 30–300 MHz wideband biconical antenna is 1.30 m long) and sensitive to reflections from nearby metal objects as well as pre-existing site RF ambient. Our temporary test site has many reflective surfaces (shelves, metal desks, chairs, lab benches) that may cause field peaks and nulls: *no dependable RE test can be run in such environment*. Instead, we will *measure some parameter that is proportional to the radiated emission, not the radiated emission itself* (Ott, [13]). Small magnetic loops, much smaller than a wavelength, can be brought very close for "sniffing" (<10 cm) to the EUT. Shielded loops are available off-the-shelf at affordable prices. As an alter-

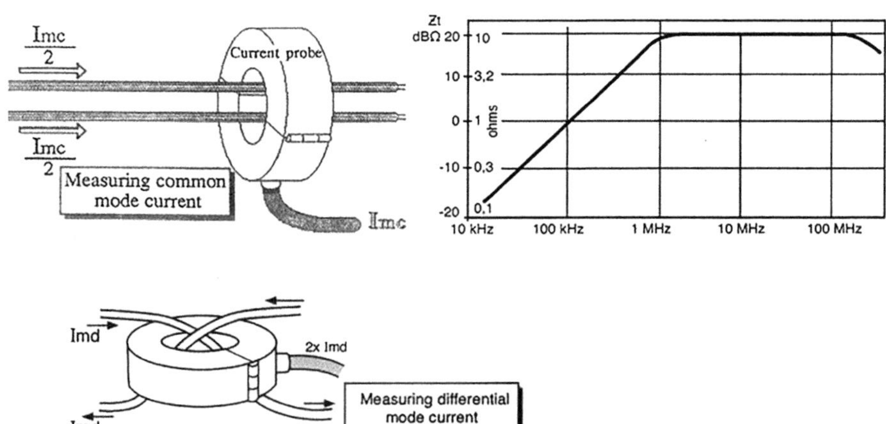

Fig. 10.2 Example of clamp-on probe, mounted for segregating CM from DM currents, and its Zt calibration curve (right). This one is useable from 10 kHz to 300 MHz

native to a commercial one, a simple homemade probe can be constructed from a 50-ohm coaxial cable (Fig. 10.3).

- ***Good quality coaxial cables.*** Experience tells that a significant amount of time and effort is often spent chasing odd and non-reproducible results caused by low quality or second-hand coaxial cables and especially coaxial connectors. The integrity of the braid and perfect, 360° contact of connector backshell and mating parts with the receptacle are important for dependable measurements. Brand-new RG58 and BNC set can do a fair job, but coaxial cables of dubious origin with worn-out BNC can ruin a series of test records. For reliable results, especially with emission tests, prefer double braid coax with N or SMA connectors (male AND female), because of their threaded instead of bayonet fittings. In addition, double-braid coaxial cable exhibit lower losses above 100 MHz.

10.3.3   Conducted Emissions (CE) on Power Cord

The majority of CE specifications are addressing only the HF noise present on the main power cable. We will see that for *RE investigation*, substantial time savings and progress can be done by measuring also the noise present on I/O cables.

Prepare in advance a coarse list of the potential HF sources and their basic frequencies. A wise test program will anticipate what type of repetitive (or eventually random, non-coherent) noise could be present on EUT cables [12]. This will facilitate the identification of BroadBand (BB) versus NarrowBand (NB) nature of the emissions (see further discussion, Peak/Quasi-Peak). With the help of its designer, list the EUT operating modes that will exercise the maximum of its internal or I/O functions, and retain the ones that are likely to cause the highest activity.

Since you are not in a Faraday cage, beware of the ambient HF noise that could corrupt your measurements. EUT cables and your own instruments can pick-up

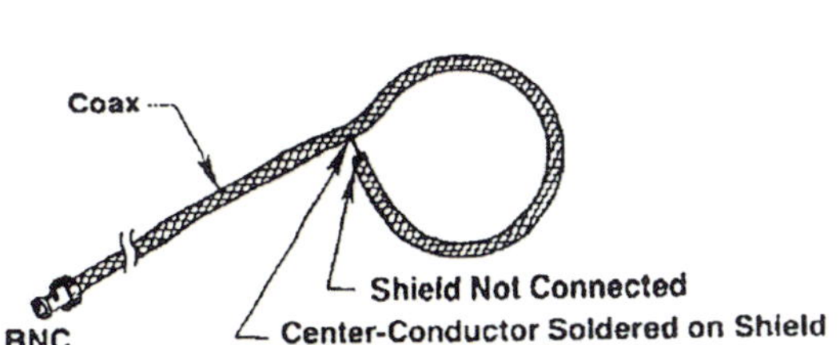

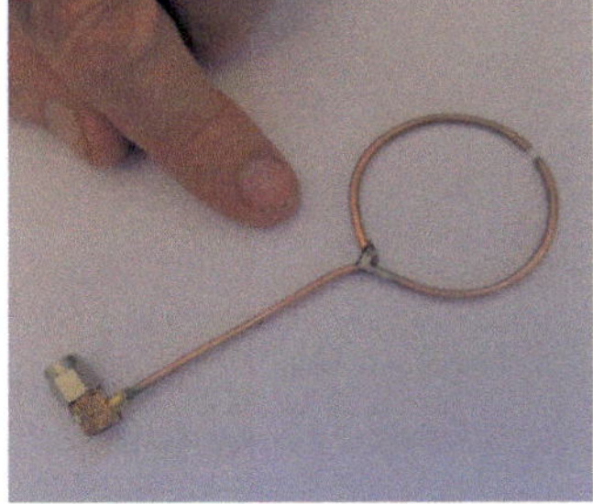

Fig. 10.3 Examples of small, home-made H-field probes. The right-hand photo shows a 4.5 cm diam. Probe made of a miniature semi-rigid coax. The shield gap has been shifted midway on the loop, which moves the self-resonance of the loop up to a higher frequency. The calibrated probe factor for this 4.5 cm loop is: K (in 50 Ω load) = 1 μA/m per μV (that is 0 dB), *flat from 70 to 700 MHz*

emissions from radio stations and other ambients. All the same, if the EUT is associated with ancillary equipment, make sure these items, if supplied by their individual power cables, are either compliant with the very CE spec limit we are looking for, or are equipped with an efficient line filter, even as an add-on. Dressing of all cables at 5 cm above the ground plane (Fig. 10.4) will also restrain their pick-up of RF ambients.

Check for ambient background noise by running several sweeps of the spectrum analyzer, with the LISN, and/or current probe in place and the *EUT turned off.* The read-out of voltage (dBμV) or current (dBμA) should be at least 6 dB below the CE limit. If this condition is not met, a CE evaluation is not feasible. However some tricks are worth trying, if there are only a few frequencies where the background noise is too high:

– Record those frequencies where the background noise exceeds our limit
– Change the scan width of your analyzer down to 50 or 100 kHz per div.

Since your receiver bandwidth for a CE below 30 MHz is 9 or 10 kHz, chances are that when you will turn the EUT "On," its signature will show out of the background spectral lines, or in-between. *This procedure is safer than turning the EUT "ON" then "Off," watching for the differences.*

10.3.4   Test-and-Fix Routine

With the EUT "On" sweep the prescribed frequency range using the **peak, maxhold function** if available, overlaying 5–10 sweeps. Keep at least 10 dB attenuation at the Sp. Analyzer input, since dynamic range is seldom a problem with CE measurements. Record the EMI voltage (dBμV) at each LISN port, and retain the worst

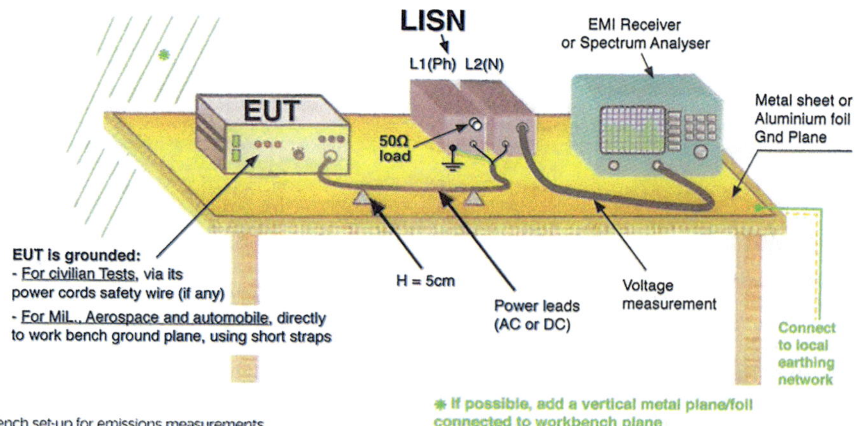

Fig. 10.4 Work bench set-up for emissions measurements

value (Phase vs. Neutral, or L_1 vs. L_2). If the limit is in dBµA, do the same with a current probe successively on L_1 then L_2 line. Make sure that the non-tested port at the LISN set is fitted with its 50 Ω load (many commercial LISNs do this automatically).

Where a limit violation appears (ΔdB), return and zoom to these specific frequencies trying to catch the culprit element by turning off, or disconnecting temporarily one of the following:

- each one of the switch-mode regulators that could generate harmonics corresponding to the limit violation,
- same for the processor card, if it can be unplugged or set on standby with EUT still "On,"
- some loads that are notoriously noisy: motors, discharge lamps, etc.

CAUTION: unless there is a surge limiter on the RF input, NEVER turn-off the EUT while the spectrum analyzer input is connected to the LISN port.

Receiver Detection Mode: Peak, Quasi-Peak or Average?

Normally, CE limits on power line are defined in a 9 kHz Bwidth and average detection. However there is a second, more liberal limit for BroadBand emissions, using the Q-Peak detector. Since it allows for faster sweeps, start with Peak detection:

(a) If the Pk detection display is compliant with Average limit, *the EUT is compliant.*
(b) If the Pk measurement exceeds the Average limit at some frequencies, you get a second chance,
(c) Repeat the measurement with the QPk detector at the sole frequencies of concern, because it obliges to slower sweeps). Compare with the QP limit, which is 10 dB more permissive:

- if the QP limit is exceeded, EUT is non-compliant
- if the QP limit is met, repeat the measurement in AVERAGE mode
- if Average Limit is met with the Aver. Detector, the *EUT is compliant* otherwise it is not

Common Mode (CM) or Differential Mode (DM)?

This is important to know for selecting the optimal fixes. DM emissions (L_1-to-L_2) are generally strong below few hundred kHz, since they correspond to the first harmonics of the switch-mode power supplies. DM emissions are also stronger when the power input setting is lowest (for instance, 115 V instead of 230 V), or more generally when the EUT draws the maximum line current; in that respect, DM signature increases if the EUT is most active, and decrease with EUT in stand-by mode.

In contrast, CM signature does not change with the EUT activity, but decreases if the power input setting is lowest. Segregating CM from DM contributors on a power line can easily be done with the current probe (Fig. 10.5).

Monitoring your progress with the spectrum analyzer, try reducing the off-spec levels by at least (ΔdB) + 6 dB margin, doing the following:

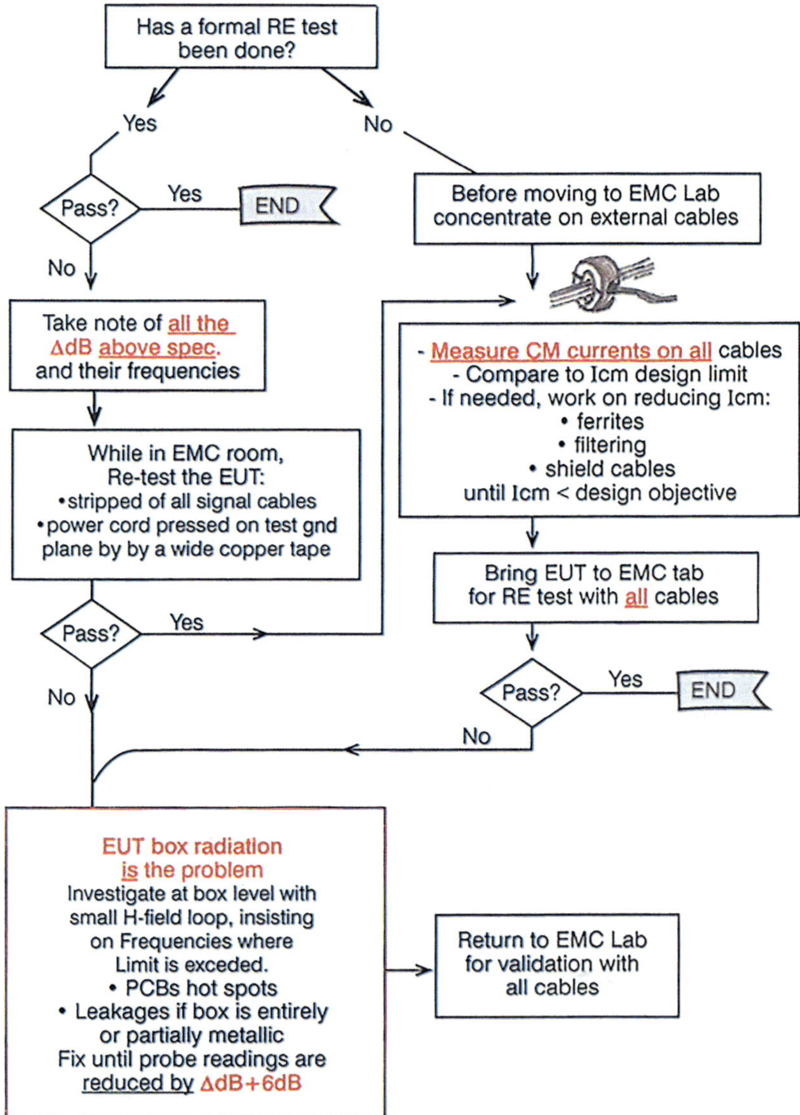

Fig. 10.5 Decision chart for troubleshooting emissions problems after a first RE test, by probing CM current and near field H loops. (From [13])

– *if the violation is by DM noise,* try a larger value for DM filtering capacitor, they are generally less bulky than magnetics. Install it between the filter choke and the power input, keeping ultrashort leads.
– *if violation is from CM noise,* look at the way the CM filtering capacitors (line to chassis) have been mounted. They are often connected with long traces or leads: a 5 nF "Y" class capacitor mounted with 3 cm total lead length (2 × 1.5 cm) becomes worthless above 12 MHz.

- Check-out the line filter schematic. Is there enough inductance for CM and/or DM reduction? Improve the filter efficiency (DM or CM, depending on your finding). If room permit in the equipment, try a filter with more efficient DM or CM choke.
- Look at the filter mounting: filters are often mounted too far inside the equipment instead of right at the power input. Watch for a possible coupling between input and output wires (or traces), and correct if needed. Then validate your progress by a formal re-test.

10.4   Radiated Emissions (RE) Check by Substitute Methods

Complying with RE limits is one major EMC challenge during product development and testing. Often, a product has been developed using the best home-grown experience plus simulation software to "make-it-work," free of internal problems. Radiated emission is one of these secondary concerns that are pushed away to the day of tests to see if it passes. Needless to say, generally it does not, unless a serious EMC analysis has been carried along design phase. The methods recommended here are time-savers for identifying and reducing quickly out-of-spec radiations, before bringing the EUT to an anechoic EMC test chamber for a true radiated emission test at 1 or 3 m distance.

While the majority of CE specifications limits are addressing only power line emissions below 30 or 50 MHz, the rationale of our investigation method is that ANY external cable, by the Common Mode (CM) current it carries, can easily radiate more than the box itself up to 200–300 MHz. This is because the mere geometric dimension of the I/O cables, usually exceeding 1 m, makes them efficient antennas, while the internal EUT's PCB traces represent dipoles or loops with sizes one or two order of magnitude smaller.

Ideally, with intentional (differential-mode) signals, the current flows down one wire of the cable, returning via a close, adjacent wire, hence the net current should be almost zero and the CM radiation be non-existent. Since an actual I/O interface is never ideal, the CM current is the unbalanced current (current not returning on the cable). If this current is not returned on the cable, where does it flows? Via the cable stray capacitance, which means radiation. Thus, measuring the undesired CM current on each I/O cable is one of the most useful things you can do (Ott, [13]).

Above $\approx$300 MHz, were wavelength λ becomes less than 1 m (i.e. $\lambda/4$ <0.25 m), the cable resembles an antenna that progressively "shrinks", while the internal EUT wiring and PCB traces eventually override cable radiation [4].

10.4.1   Measuring CM Currents on Cables

The CM current can easily be measured with a calibrated high-frequency clamp-on probe and a spectrum analyzer as shown in Fig. 10.5. The set-up remains the same as for the true CE test on power line, but this time *ALL cables will be measured.* Current probe must be moved along the 1.50 m cable section that is closer to the EUT box to make sure you do not miss a maximum of current standing wave.

Based on simple antennas formulas [12] Table 10.2 provides the maximum CM current that can be tolerated, from 30 to 400 MHz, on I/O cables for civilian residential (Class B), industrial (class A) and Military (RE 102) compliance. Our Pass/Fail criteria are based on the following assumptions:

- *for civilian FCC/CISPR,*

  - frequencies >30 MHz
  - cable length greater than 1.50 m
  - cable height: h $\geq$ 0.75 m A 5 dB margin has been accounted for ground reflection

- *for Mil 461F-RE102,* additional factors are coming into play:

  - Cables are laying 5 cm above the ground plane
  - The limit relaxes progressively above 100 MHz (Mil 461E, F)
  - Actual field measurements are made at 1 m distance

Notice that, up to 75 MHz and *regarding cable radiation only*, the criteria for the most severe Mil.Std 461 limit are quite close to what would be required for FCC-15 or CISPR 32 class B. If each I/O cables satisfy the I_{cm} table limits on every spectral line, we know that, at least, the contribution of the cables will keep us below the spec limit. *If we fail this CM current test, we will surely fail on the Radiated Emission test.*

A *word of caution*: not being in a RF-clean environment, your cables may pick-up from external sources such as local FM and TV broadcast stations. Therefore all

Table 10.2 Maximum allowable CM current on external cables for RE compliance

		50 to 230 MHz	230 to 400 MHz
· for CISPR Class A :	I_{cm}	10 µA (20dBµA)	20 µA (26dBµA)
· for CISPR Class B :	I_{cm}	3 µA (10dBµA)	6 µA (16dBµA)

		50	75	100	200	300
	F(MHz)					
· for **Mil. 461-RE102**	E_{Limit} (dBµV/m)	24	24	24	30	34
most severe limit (Air Force)	I_{cm} **(dBµA):**	10	7	4	4	4

measurements must, be validated to assure that you are measuring what you think you are measuring. A simple check is to turn the product Off and see if the reading goes away. If it stays, it is due to external pickup. For instance, signals in the 88–108 MHz (FM) frequency range should be suspect, and double-checked by turning the EUT "Off."

Numerical Example

Using the current probe shown in Fig. 10.5, the following maximum voltage values have been recorded on the Spectr. Analyzer (50 Ω input). Once translated into cable current, do we meet the criteria for class B radiated emissions?

Freq:	50 MHz	80MHz	250MHz
1) Probe read-out V(dBuV)	38	46	34
2) Probe factor (dBΩ)	20	20	16
3) Actual current I(dBμA) = (1) - (2)	18	26	18
Our Class B criteria for I_{cm}:	10	10	16
Anticipated ΔdB off spec:	**+8**	**+16**	**+2**

The I_{cm} limit is exceeded by 8 and 16 dB at 50 and 80 MHz, leaving no chance for a real RE test. There is a sleek chance that the 250 MHz emission be OK, but the margin will be thin.

Warning The I_{cm} limits are peak reading, as displayed by the Sp. Analyzer. Make sure that they have been measured with a 100 or 120 kHz BWidth, with a video BWidth set at 0.3 or 1 MHz. For instance, if some of the above-limit lines are harmonics of a 30 or 50 kHz switcher, there will be 3 or 2 harmonics adding-up in a 100–120 kHz bandwidth, compared to what has been seen in the conducted emission (CE).

This technique works on shielded cables, too. It can be a good way to pin-point possible weaknesses of a cable shield, including its terminations. A question often arises: "If our probe sees a current on the cable shield, is it really the shield current that we measure?". Typical answers are "yes".... NO! It is the CM current that escapes the shield and returns by the outer, invisible path, which means radiation. This gives a measure of the quality of the shield, whatever it is.

10.4.2 Interpreting and Reducing CM Emissions from I/O Cables

Once the I/O cables' contribution has been measured, if they violate Table 10.2 criteria—as they often do on prototype or pre-production items, we will work at bringing them below the limit.

While trying improvements, measure one cable at a time with the current probe. If the EUT is still amenable to limited changes in the PCB I/O ports areas, and/or Power input area, use surface-mount CM ferrites, surface-mount signal filters or filtered connector sockets, according to the function of the faulty cable.

- If the faulty cable (or one of them) is the power cable, don't take for granted that the filter is not guilty just because you passed the formal CE test. Power line filters are often optimized up to 30–50 MHz, because this is as far as the CE spec goes, and their attenuation could very well drop beyond this frequency range. Try to install an additional bifilar-wound ferrite, or small ceramic CM capacitors, line-to-chassis, close to the power entry port.
- Check that there is a good, metal-to-metal bonding of the PCB Zero Volt (signal Gnd) to chassis, close to every I/O port.
- Inspect the internal EUT wiring, checking for possible crosstalk between I/O wires and internal wire/traces that do not come out.
- Replace unshielded cables/pairs by shielded ones, or if already shielded, check that the shields are making a perfect (360°) contact with their receptacle, straight to the EUT chassis.

After each fix or set of fixes, repeat the Spectrum scans with the current probe to check if—and by how many dB—you have reduced the CM current, until NOT ONE spectral line exceeds our CM current limit. Once you have been through all cables one by one, run an overall check of the CM currents: some may have increased on a previously fixed cable (the classic "balloon effect").

At this point you are confident that the cables will no longer cause a violation of the radiated emission test at a qualified EMC facility. However, you are left with a last risk: chances are, especially with fast logic with clock speeds >30 MHz, that the EUT box itself radiates above the permitted limit.

This is where a miniature near-field probe can be used. Without trying to convert the readout into E-field, crude measurements made with such H-field probe allow making A/B comparisons of the Sp. Analyzer scans, when improvements are attempted. This implies that we have at least one actual RE measurement for the EUT box-only radiation on a qualified test site, as we will see next.

10.4.3   Measuring Radiated Emissions from EUT Box Alone (I/O Cables Excluded)

Once the I/O cables contribution has been checked and reduced, an actual RE test in an EMC test lab can be made, with fair chances of success, except for the risk of radiation emanating from the EUT box itself. If the EUT with its EMI-hardened I/O cables still fail the test, we will re-test without the I/O cables (Fig. 10.6).

Near Field Measurements on PCBs

What we can do first is to identify the strong magnetic fields close to the printed circuit board using a small magnetic field loop probe and the spectrum analyzer. The magnetic field probe is held in the three axis (X, Y, Z) when performing the tests, such as to capture the maximum field strength. Because such small loops are insensitive to remote external fields, you can assume that what you get is coming from the nearby PCB. This can be validated by moving the probe 2 or 3 times further away from the board: the reading should drop abruptly by 4–9 times (in near-field, H-field is falling-off strongly with distance. Because of this, **make sure to keep a constant distance from the center of the loop to the PCB zone you are suspecting**. A simple way is to stick an insulating spacer used as a distance gage (Fig. 10.3).

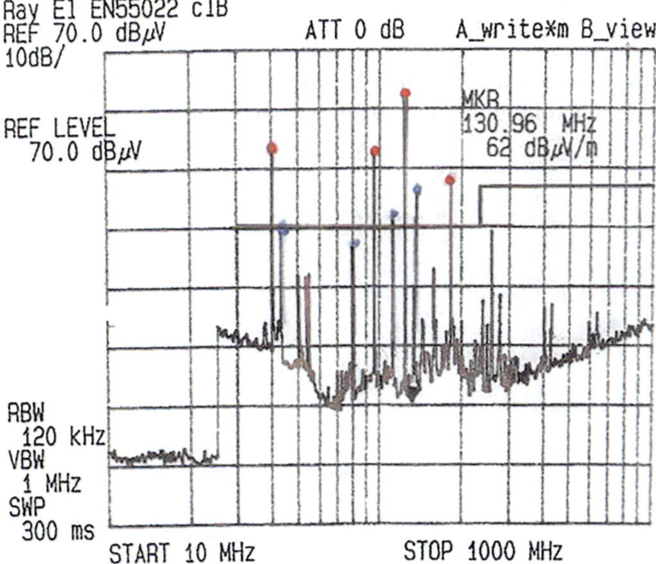

Fig. 10.6 RE test results and investigations. Red dots represent the contribution of I/O cables to the limit violations, which disappear when cables were removed. Blue dots are the off-spec (or close to) radiations due to EUT box alone, that will be investigated and fixed separately

Scan the probe over the printed circuit board looking for "hot spots" (locations of strong magnetic fields). When some are found, check the PCB in that vicinity for violations of good EMC design practices. One frequently found is an interrupted signal current return path caused by a split or slot in the ground/power plane [2]. If a quick, temporary change can be made to the board, retest to confirm that the H-field has decreased in amplitude. You may often find that it is an IC module that is causing most of the emission. In this case, consider using a small board-level shield over the component(s) or a small ferrite "tile" on top [12] (Fig. 10.7).

Near Field Measurements Around the EUT Box

If the EUT is housed in a—supposedly—shielded enclosure, with little hope that the PCB could be modified, try to identify electromagnetic field leakages through the apertures, using the H-field probe [3]. Place the probe close to the enclosure with the plane of the loop in the three axes (X, Y, Z). *Keeping a constant distance*, move the probe along the seams, apertures or connectors areas and search for a strong magnetic field. After making changes to the enclosure (e.g., reduce the aperture size or length of the seam, add more screws or spring contacts, temporarily cover the leakage with large conductive adhesive copper tape, etc.), retest to confirm that the magnetic field has decreased in amplitude.

For both PCB and Box leakages treatments, make sure that your H-field probe read-out after the fix has been reduced by at least (ΔdB + 6 dB), Δ being the amount of your limit violation.

Note In theory more could be done if the H-field probe is properly calibrated. Unfortunately, calibration curves provided by the manufacturers are a nightmare to the user, given in exotic units like "dBm per microTesla," requiring some legwork

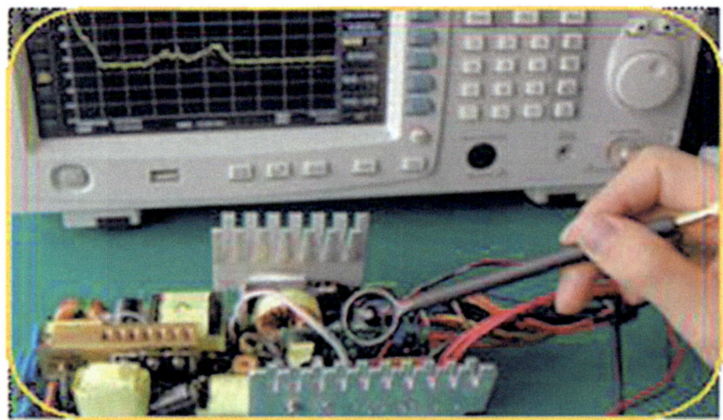

Fig. 10.7 Identifying radiated "hot spots" on a switch-mode power supply PC board. (Courtesy of RIGOL)

for a quick translation in dBµV per dBµA/m, and conversion into an equivalent E-field at 1 or 3 m. This includes near field-to-far field conversion factors, a computation that we do not recommend to other than seasoned, full-time EMC experts.

10.5   Part II: Troubleshooting EMI Susceptibility

As aforesaid (Sect. 10.2), conducted tests are faster and easier to perform. So only a limited set of susceptibility tests can be carried using essentially conducted excitation set-ups, without (or before) resorting to an actual Radiated Susceptibility check in an EMC lab.

Preferably, susceptibility checks should be done after emission tests have been carried-out, and eventually their non-compliance has been solved. Notice that so-called *Susceptibility* tests are in fact made to demonstrate *EUT's Immunity*. Therefore, we should always try to reach, within a reasonable time span, the actual *"Fail" threshold* to get a measure of our *Compatibility margin* as compared to the Immunity "No-Fail" requirements.

10.5.1   Troubleshooting Susceptibility on a Prototype or Pre-qualification Item

Like for emissions, we try staying away from formal Radiated Susceptibility (R.S) tests and use substitute conducted tests instead. These conduction (or injection) tests are deemed to recreate on the EUT what its cables (power and signal) would actually receive from a severe, conducted or radiated threat. So, by the order of simplicity and efficiency, we will limit our tests to the conducted tests described in Chap. 9 and Fig. 9.10:

– Electrical Fast Transients (EFT Bursts)
– ESD (Direct and Indirect).
– Bulk Current Injection, Continuous Wave.

10.5.2   Requirements for Informal Susceptibility Test Site and EUT Set-Up

The site requirements for an informal EMC Immunity test are the same as for emissions (Sect. 10.3) except that the quiet *RF ambient requirement is exactly the opposite*: a quiet ambient is not necessary. Instead, since susceptibility tests will force strong HF signals or fast pulses, it is prudent to forbid or temporary discontinue the use of sensitive electronic equipment (office equipment, computers, and instruments other than intentional EMI generators) in close proximity—say less than 5 m from the test set-up. For the same mirror reasons as above, it is recommended to filter and if possible isolate the AC power branch circuit that feeds the EUT and set-up. This can be done by the same LISN and isolation transformer that were used for emission testing.

During the susceptibility test, we will limit the Auxiliary Equipment (AE) to a minimum, using everywhere possible passive loads at the end of I/O cables. A good example is a Tx/Rx link that can be set in self-looping mode. As for emissions tests, the *AEs must have an Immunity level equal (or greater) to the one we are testing for.* If this is not feasible, a temporary filtering on their interfaces or Coupling/ Decoupling Network (CDNs) can be used to isolate the AEs from the EUT.

10.5.3   EUT Preparation for the Test

Bear in mind that you will be "teasing" a product to see if and how it fails. To easy-up the test, the EUT must be set in an automatic mode such as it performs its normal functions continuously. This can be done using an auto-test or a specific emulation software which:

(a) Constantly activates a real or close-to-real application w/o operator intervention.
(b) Indicates clearly a disability or loss of performance by, like:

- displaying errors
- printing a report
- activating a visual or audio alarm (LED, buzzer, etc.)
- reset or lock-up in a frozen state

All this without the need of external instrumentation/hook-up probes, unless they are fiber-optic links. *This is paramount to a meaningful test.*

10.6   Testing for Immunity to Electrical Fast Transients

Of all susceptibility tests, the EFT (or "Bursts") is the simplest to install and run, and yet one of the most meaningful. Thanks to its wideband spectrum covering kilohertz to at least 60 MHz at each single pulse, it will track down most of the EMC weaknesses of digital as well as analog circuits.

Logic functions can:

- mistake the envelope of the bursts for a valid bits train
- have edge-sensitive gate inputs triggered by individual short glitches
- suffer clock stretching if the device fails to receive ACKs from the other device
- loose signal integrity due to high noise on DC supply and Gnd to which communication I/Os are referenced, violating the protocol specification

With analog circuits, the envelope of the bursts can be detected in a same manner as pulse modulated radio signals.

Because it simulates the direct or indirect coupling of switching transients (relays, repetitive arcing in switches, turn-off of heavy-duty motors, variable frequency converters, etc.) caused by nearby power devices on the AC distribution wiring, the test is applied both on the power input and signal cables, per IEC 61000-4-4.

10.6.1 EFT Instrumentation and Set-Up

EFT generator and EUT should be installed on the test ground plane (floor level or bench with metal foil), with cables laid at approximately 5 cm above with insulating spacers (Fig. 10.8). *The generator should not be too close from the EUT*, since 4 kV pulses with 5 ns rise times will radiate a strong E-field. Close exposure of the EUT could cause a direct box-to-box failure, misleading the test issue. Same remark applies to the associated AE.

- For superimposing HF pulses on power cord, the generator has its own internal Coupling/Decoupling network, such as the fast bursts are injected on the EUT power input only and do not sneak into the lab power mains. Injection is normally CM (Line 1 + Line 2 together vs. chassis), but it is possible to select any combination of L1, L2 and earth (Green/Yellow).
- EFT injection on I/O cables is a *contactless coupling* (no stripping of cable insulation), applied with a special 1 m long capacitive clamp, folded over the tested cable (Fig. 10.2). This folding armature is the *coupling "hot plate," reaching up to 4 kV and must not be touched or approached during the test*. Since its size and weight can be impractical for some non-standard set-ups, it can be replaced by a tight wrap of aluminum foil, 30–50 cm wide, around the tested cable, that will provide the necessary 60–100 pF of distributed capacitance to the tested wires. The HV injection cable from the generator should be ≤1 m long.

10.6.2 EFT Test Routine

For all injections, test at least 30 sec. on each polarity. Do not switch abruptly from (+) to (−) pulse polarity, always step through an "Off" mode in between.

Consider this as a **no-damage // no errors criteria:**

The low amount of energy in each pulse is such that a damage level is seldom reached without a preliminary warning by a malfunction or error. DO NOT insist in reaching a true damage level! Keep your prototype safe: investigate the reasons of the temporary fault—or lock-up status.

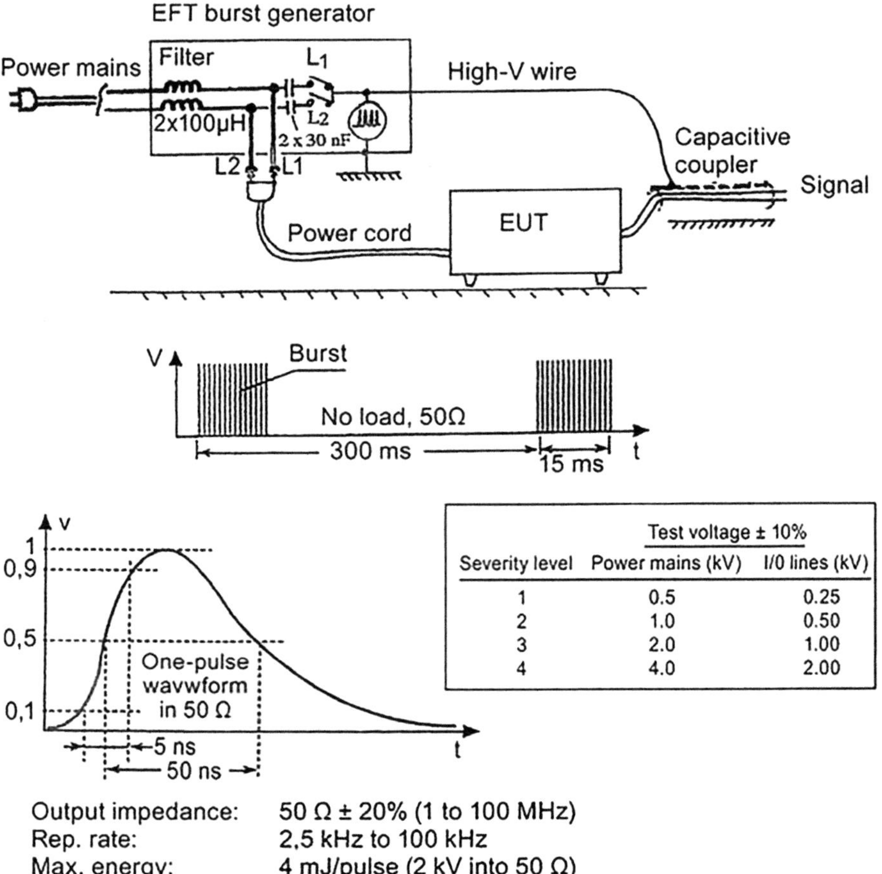

Severity level	Test voltage ± 10%	
	Power mains (kV)	I/0 lines (kV)
1	0.5	0.25
2	1.0	0.50
3	2.0	1.00
4	4.0	2.00

Output impedance: 50 Ω ± 20% (1 to 100 MHz)
Rep. rate: 2,5 kHz to 100 kHz
Max. energy: 4 mJ/pulse (2 kV into 50 Ω)

Fig. 10.8 EFT/surge generator and test set-up, for both standard and on-site conditions. (Courtesy of EMC Partner)

(a) If nothing else specified, start with the lower level of IEC 61000-4-4, then increase progressively up to at least a No-Fail level #3 (2 kV) which is generally satisfactory. However, for a development test, it is advisable to try reaching one level higher to know your margin.

(b) *Increase the level progressively*, preferably by octave steps, addressing all combinations of (+), (−), L1, L2, L1 + L2 and so on. When (if) a FAIL level is reached, note the settings for all the last RUN levels. If the generator has progressive voltage setting, reduce the amplitude until you find the actual RUN/ FAIL edge, such as you know your margin.

(c) *Next, inject onto each signal cable*, keeping the capacitive coupling sleeve 1 m away from the EUT. If external cables are shielded, the test pulse is simply applied to the shield insulation jacket. If an AE was needed for the test monitoring, make sure that it has its input protected by filtering or snap-on ferrites toroids, located near the AE side.

10.6.3   EFT Test Diagnosis and Fixes

If the test fails, concentrate on the failing port (Power or Signal) and check for:

– Incorrect mounting, or insufficient CM attenuation of the powerline filter
– Missing or poor cable shield connections
– Ungrounded metallic connectors
– Lack of, or improperly mounted CM filtering capacitors (long leads)
– I/O wiring running too far inside the EUT before they are filtered

Try improving immunity by correcting the above, eventually adding filtered connector/adapters, or split-ferrites (two or three turns) to the faulty I/O cable.

Restart the test, starting from the latest "NO FAIL" level as a reference, for grading the improvements.

Warning If a fix does not seem to bring a significant improvement, do not remove it. Add another one: EMI is often a multipath scenario.

10.7   Electro-Static Discharge (ESD) Test

Next to EFT, the ESD test is also a powerful teller of potential EUT weaknesses. It covers in a single flash an even wider frequency spectrum, up to 350 MHz considering its 0.7–1 ns risetime, but at difference with the EFT conducted bursts, it generates also a strong and fast rising H-field on the EUT box itself. You are directly zapping in a few ns a broad frequency spectrum that would take hours to explore during a real RF immunity test.

10.7.1   ESD Instrumentation and Test Set-Up

The following will be needed:

• An ESD generator conforming to IEC 61000-4-2, with 330 Ω internal resistance and 150 pF cap. is the most common instrument. Other simulators can be used according to the family of EUT being tested.
• Horizontal and vertical coupling plates (HCP, VCP) if the EUT must be tested for indirect discharge (IESD).

EUT preparation and set-up precautions are the same as for other immunity tests, with a strong emphasis on proper decoupling or safe distance of the associated AE, if any. The location of the ESD current ground return lead is critical: it should be attached to the test ground (floor) plane on the same EUT side as the face being tested (Figs. 10.9, 10.10, and 10.11).

10.7.2 ESD Test Routine

At difference with the continuous train of high-voltage bursts, ESD replicates an isolated, rarely-occurring event. Therefore, exceptions accounted, there is no point in applying an "absolutely-NO error" criteria. If the EUT has efficient error-detection and recovery mode, this can be considered as an acceptable response, provided it does not result in a lock-up status, requiring a manual reset or creating a safety issue.

(a) Establish the Direct or Indirect (IESD) procedure you will be using (ref. IEC 61000-4-2).

Briefly, if the EUT is metallic on all its faces, only direct ESD will be performed. Otherwise direct discharge will be attempted on all touchable or arc-reachable metal parts, the rest of the EUT being tested by the IESD method on the HCP and VCP. Insist on the cable entry areas, keyboard and display edges, and touch sensitive panels. Mark-up the pre-determined discharge points directly on the EUT faces or on 3D sketch.

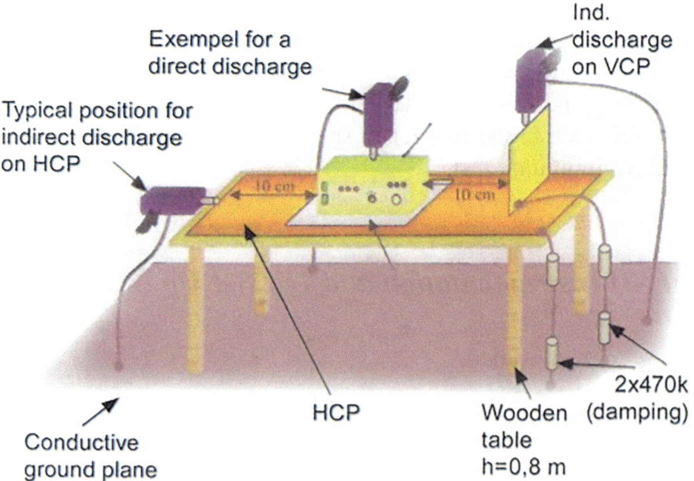

Fig. 10.9 ESD test set-up, for table-top equipment. The horizontal (HCP) and vertical (VCP) coupling plates are used to simulate a discharge on nearby structure, especially when the EUT is non-metallic

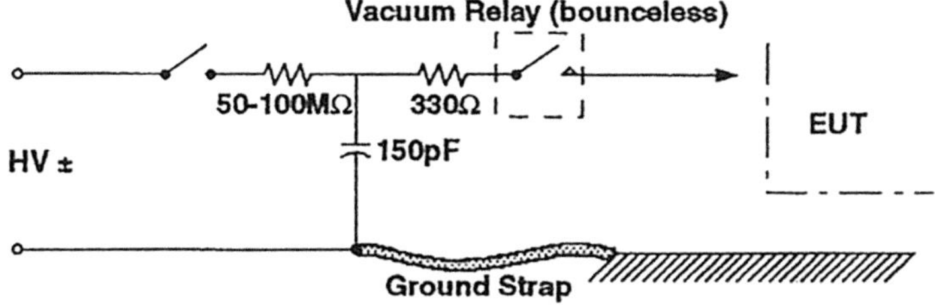

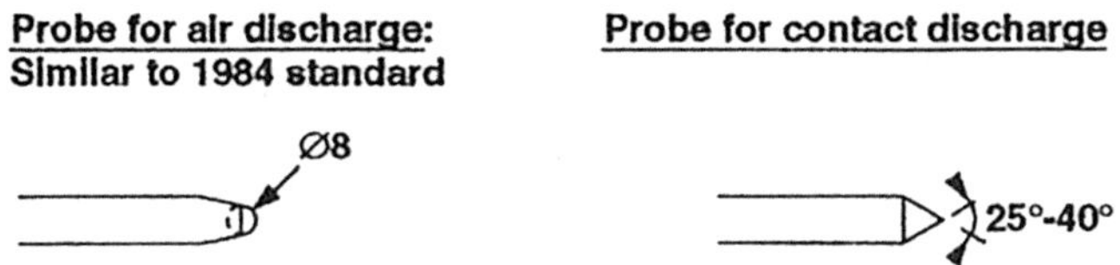

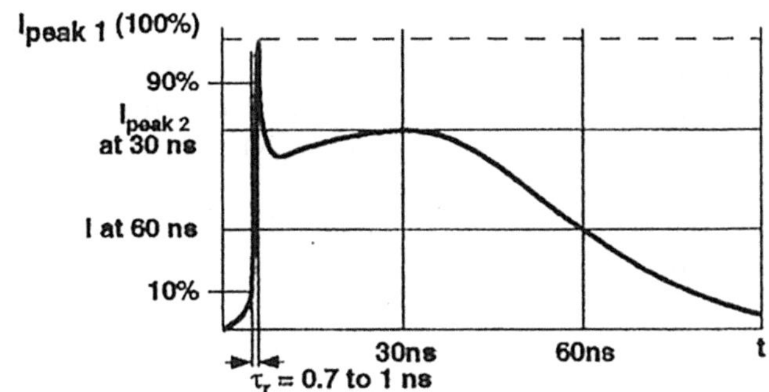

Level (Contact Discharge)	HV (kV)	I_{peak1} (± 10%)	I_{peak2} at 30ns (± 30%)	I at 60ns (± 30%)
1	2	7.5A	4A	2A
2	4	15A	8A	4A
3	6	22.5A	12A	6A
4	8	30A	16A	8A

Fig. 10.10 ESD simulator offering several discharge heads. Simulator setting on severity #1 is an absolute minimum. For equipment in ESD-prone areas, aim for severity 3 or 4. Highest severity for air discharge (no contact) is 15 kV. Charging/discharging circuit is generally packaged in a compact, hand-held "gun" head. (Courtesy EMC Partner)

(b) Start with a low level (e.g., 2 kV) in a fast repetitive mode, like 5 or 20 pulses/s for a coarse survey of all the discharge zones.

(c) Increase the level progressively by 2 kV steps until you reach the desired criterion plus a 1–2 kV margin.

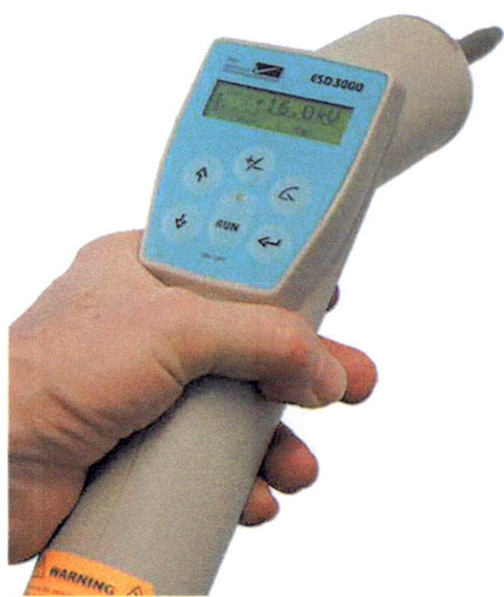

Fig. 10.10   (continued)

(d) If failures are noted, abandon the fast repetitive mode and return on the failing area with a single shot or <1 pulse/s mode, such as allowing for self-recovery process, if any. *Run a minimum of 50 discharges per point.* ESD is a super-short duration, random event: applying only a few discharges (the ten discharges recommended in IEC norm are definitely not enough) would turn into a risky hit-or-miss process [16, 17, 19, 20].

(e) Record the Run/Fail levels, documenting the type of error/malfunction: Recoverable? Non-recoverable? Loss of data?

10.7.3   ESD Test Diagnosis and Fixes

If the test fails, try to understand what are the sensitive circuitry that have been exposed to the very close coupling of the discharge. Watch for long slots or seams leakages on the EUT enclosure, that act as re-radiating antennas to the nearby wiring or PCB traces inside.

When trying fixes such as:

- Shielding (or shielding improvement) of I/O cables
- Shielded connectors, ferrites, decoupling capacitors
- Better bonding of panels, doors, etc. to the main chassis (for metal cabinets or racks)
- Reduction of slots and seams leakage by gasketting or adding more screws
- Increasing the air gap crossing distance for zones where air discharges have occurred

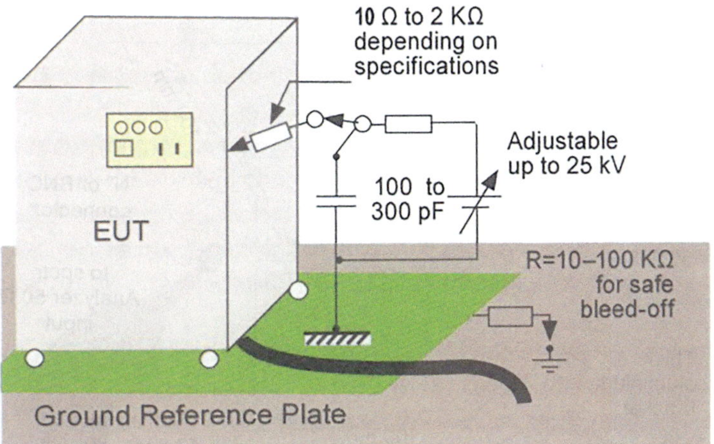

Fig. 10.11 ESD generator and test set-up for floor-standing EUT or on-site evaluation

Always grade your progress via new Run/Fail mappings. Like for EFT, *do not remove a fix that gave disappointing results. Accumulate them until success*, then, and only then, remove fixes sequentially to find the ones that could be useless.

One ultimate tentative would be to make a non-contact discharge close to the PCB which seems to be involved in the failure. It is addressing the vulnerability of the stand-alone PCB, removed from the rest of the EUT and installed on the work-bench metal plane. This is a delicate method [16] requiring a specific set-up and a seasoned EMC expert, and will not be described here.

10.8   Bulk Injection of HF Current on External Cables (BCI Test)

Next to EFT and ESD, this test is a powerful indicator of EUT susceptibility to strong radiated RF ambients, from intentional radio transmitters. It consist in inject-ing with a clamp-on current transformer *the same HF signal as the cables would see if they were exposed to a real field*. The difference being that we do not need to generate the field. This test is required by Mil Std 461 CS114, with values of injected HF current ranging from 15 to 300 mA, and to some extent, is also part of IEC immunity test 61000-4-6.

Unfortunately, running this test makes your EUT cables radiate a significant RF field, which could make you liable for such an illegal interference to public radio and TV as far as km away. We do not recommend trying it elsewhere than in a shielded room, or at least an underground place (Figs. 10.12 and 10.13).

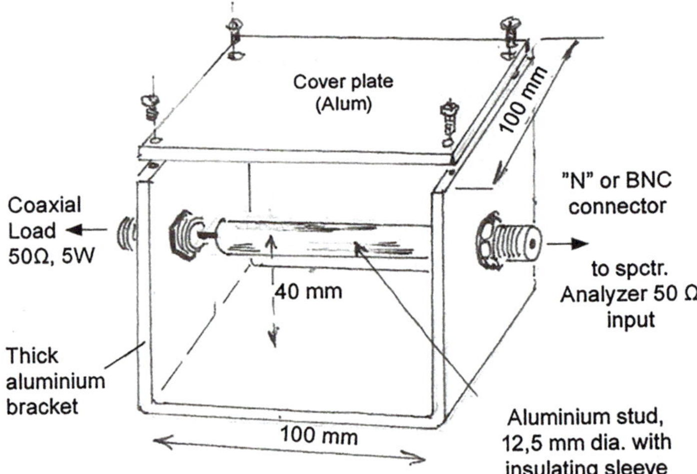

Fig. 10.12 Home-made BCI calibration jig. The aluminum center stud should NOT contact the metal frame at its ends. With the injection clamp in place, the injected current is known by a measure of the voltage on the 50 Ω output port: $I = V_{(50\Omega)}/50$. It is a substitute to an actual illumination of the entire system cables by very strong fields. Correlation is ≈ 1.5 mA per V/m

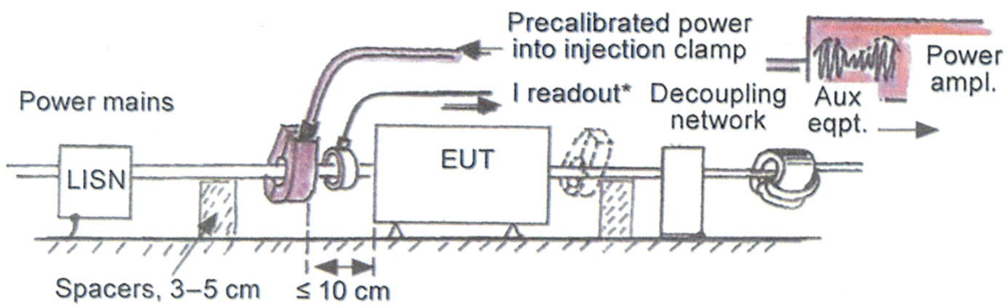

Fig. 10.13 Investigation/prequalification BCI set-up. It substitutes to an actual illumination of the system by a field. A current tracking is not absolutely necessary, if pre-calibration levels are used, but it provides the actual fail/No-Fail levels for documenting the immunity (From [12])

10.9   EMC Hardening for Equipment Installed at Customer Site

EMI problems occurring in the field are a difficult challenge, since there is very little possibility, if at all, for modifying the equipment. All solutions are relying on cable layout, shielding, additional I/O and power filtering, power conditioning and eventually room or site treatment. Even if, as of today, most electrical/electronic equipment put on market have to meet EMC requirements, susceptibility, and emissions problems still happen, for several reasons:

– a recent, large system could include older equipment that do not comply with current EMC norms
– poor cabling and grounding practices by unaware contractors, degrading an otherwise correct immunity
– some sensitive equipment are impossible to harden, by their very nature (magnetic sensors, E-Beam microscopes) or generate strong disturbances (elevators, arc welders, RF medical imaging, some industrial process)

The following basic guidelines will save you from frustrating experiences at customer's site, where people telling you that they have tried everything rely on your magic wand for making miracles.

10.9.1   *Before Going to the Site*

In many cases, we have found that rushing to a problem site because the user is pressing you may induce two mistakes:

- You did not collect enough data for preparing a sound investigation scheme.
- You either bring too many or too few instruments or measuring devices, leaving behind the one precious accessory that cannot be found on site.

Instead, there are things you must check in advance ("Forewarned is forearmed"):

(a) *Interview the users*

- Is the problem continuous or intermittent? in the latter case, is it predictable, like being re-created on request? This may dictate the equipment you will bring.
- Does the failure correlate with a specific time of day? Or with certain loads being On/Off the power distribution? When people use portable transmitters?

(b) *Is there some instrumentation available on site*? Oscilloscopes: which Bandwidth? Battery-operated? Recorders for intermittent problems? Thus you will bring whatever is necessary to complement the instrumentation available on site. **In any case, don't forget bringing an EMC clamp-on current probe.**

(c) *Sometimes, a fast, rough estimate* can be done based on the above, on the likelihood of the problem. The data you have collected may suggest what is the source, its frequency range, what and where is the victim device and what are the coupling path the EMI could take.

(d) *Plan some advance strategy for diagnosis and fix.* Don't count that the first attempt will work, anticipate some alternate options.

10.9.2 Upon Arriving on the Site

Perform a first Inspection of the EUT and its physical Installation:

(a) *On the faulty equipment itself*: Does it have power line filters, with CM and DM elements? Are they bonded to a chassis or a grounded metal plate, and correctly mounted (no input and output wires entangled together)?
(b) *General Power grounding/Earthing scheme*: Ask the customer's staff or facility maintenance people about it. Don't take anything for granted: bonding points that seem to make perfect contacts don't; Grounding straps that people swear have been installed are missing or loose; Asked about the neutral earthing scheme, six different people will give you six different versions, whereas when you check you will find that none of them is true.
(c) *Power and signal interconnecting cables layout.* Examine all of them. Are there some power and signal bundles running tightly close to each other, or near to other power cables carrying large currents? Are they shielded? If so, where and how are the shields grounded? Do they run in metal raceways, with continuous meta-to-metal bonding all along?

10.9.3 Continuous or Quasi-Continuous EMI Problems

By this, we mean that the problem is always present or occurs repeatedly within a reasonable observation period, such as it can be probed and quantified: Signal/Noise degradation, degraded Bit/Error rate, false alarms, etc. If the EUT has been checked in good condition, a continuously occurring problem is a blessing, since it eases-up the search for the source and the subsequent fixes, using the procedure shown in Fig. 10.14.

10.9.4 Measuring CM Currents as "Witnesses" of the EMI Paths

Once (if) the source has been identified, the likely coupling path(s) can be identified by a quick measurement of the HF currents at the following locations:

- power input wiring (CM)
- signal cables (CM)
- earth wires

Make sure that the identified culprit source is functioning when making your measurement. An oscilloscope with sufficient bandwidth (≥ 100 MHz) is recommended, using eventually its memory function if the culprit source is operating with

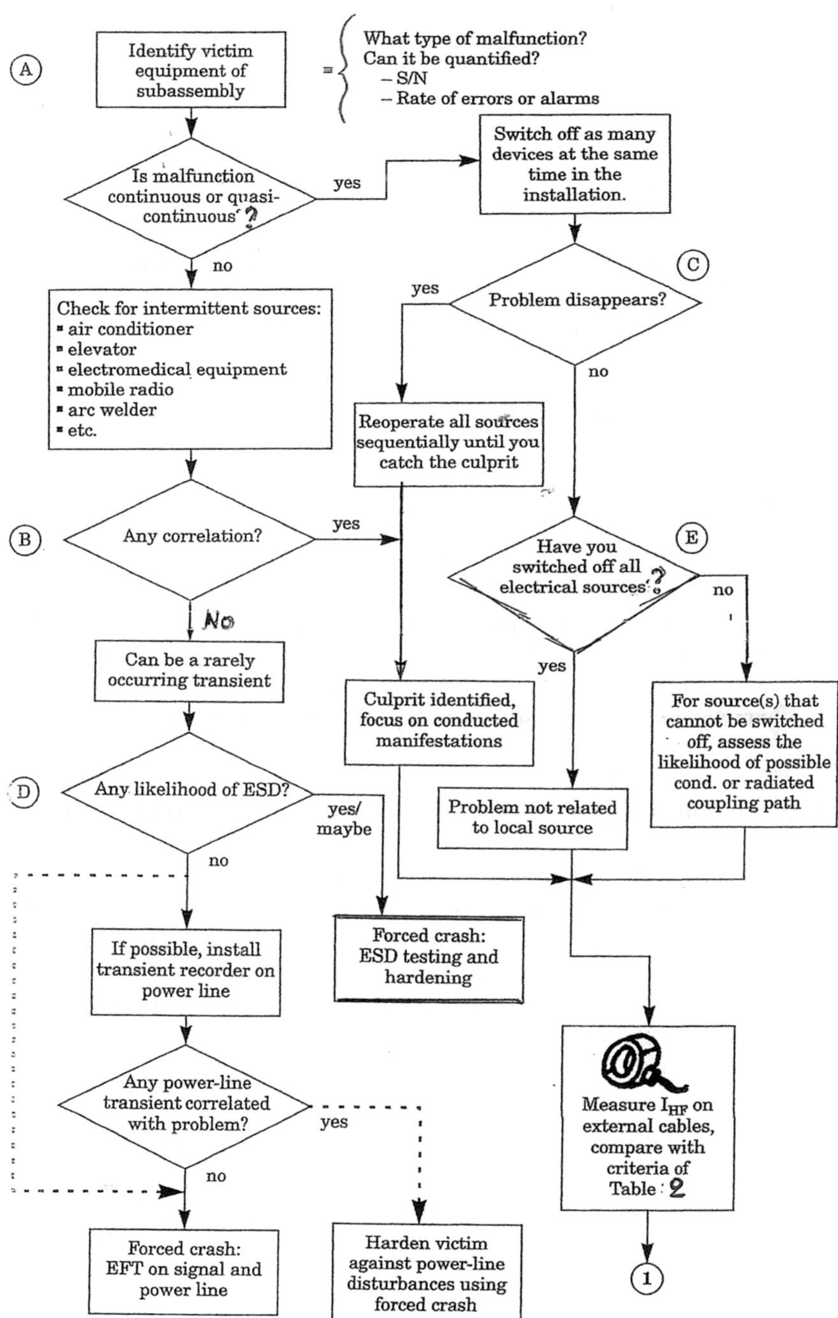

Fig. 10.14 EMI troubleshooting routine for on-site investigations. If malfunction is quasi-continuous, a measurement of HF current on cables is needed, leading to a further investigation: Exit "1" [12]

various sequences. *Make sure to use a 50 Ω input for the scope.* Using the current probe transfer impedance Zt, convert your voltage readings into current, using:

$$I(\text{Amp}) = V / Zt(\Omega)$$

Warning   the probe response (Zt) is not always flat with frequency, so you need at least a rough idea of the frequency of the signal (Table 10.3).

Upon the results of this CM current survey on all cables, a further investigation can lead to identify the cable(s) that is the likely coupling path for the problem. Figure 10.15 provides the guidance for treating such paths.

10.9.5   Intermittent Problems

These represent a respectable % of field calls, with a long search to find a correlation between a rarely-occurring problem and some external cause. If the source is found within the customer premises, it may be worthwhile seeing if the noise can be suppressed at its source, by adding filters, or box/cable shielding. Sometimes after a patient search, the external cause is found but you have no control on it (elevator in the hallway, an arc-welder in a machine shop one block away, etc.). Ultimately, no correlation has been found. These are cases where we will use the forced-crash approach, recreating artificially an event that we cannot afford waiting for.

Failures for Which No Correlation Was Found: "Forced Crash" Approach

If answer to Box "B" routine Fig. 10.14 is NO, no correlation has been found between the failure and some intermittent event on the site. Such events can be artificial or natural, occurring at irregular times in this site, like severe electrical fault triggering overcurrent protections or ground fault detectors (GFD), or eventually storms. In such case, searching in the history of local lightning strokes, or power disruption in some part of the building can provide clues if they correlate with the moment of the failures.

Table 10.3 Criteria for assessing a "Sure Immunity" and "Sure Susceptibility" for CM current readings, from $\approx$10 kHz to 100 MHz

	No problem	**High chance of problem**
HF current on power mains cables:	< 10mA @ worst frequ.	≥ 300mA
HF current on signal cables , digital:	3mA	≥ 100mA (300mA if shielded)
" " low-level analog :	0.3mA	
HF current on earthig/ground wires (in spite of fault protection installed)	< 1A	> 30A peak (transient)

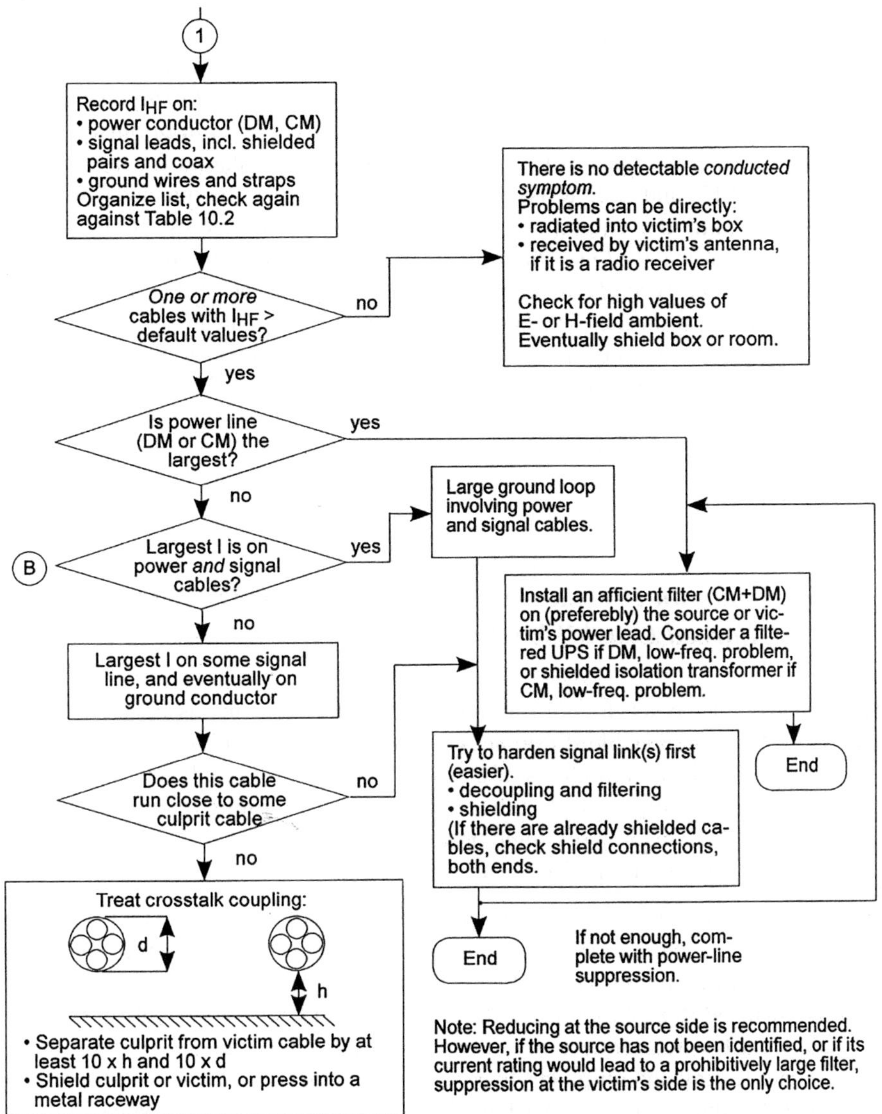

Fig. 10.15 EMI troubleshooting routine continued from Fig. 10.14, when malfunction is quasi-continuous and measurement of HF current on cables has been made [12]

Although more difficult to track, a second random cause, may be the normal ON/OFF switching of low power electromechanical switches, relays, thermostats, etc. in close vicinity (less than 10 m) from the victim.

Finally a third threat, ESD, has generally a seasonal relationship, increasing during cold and dry weather, aggravated by high personnel activity and type of floor covering. But it may be also machine-originated ESD, with production tools, robots, etc., generating ES charging/discharging phenomena with peak currents exceeding 50 A, that may go un-noticed.

In all the above cases the "Forced Crash" is an ultimate strategy where *you decide that you cannot wait for some hard-to-catch problem to show-up, so you deliberately inject onto the equipment a transient pulse (or train of pulses) that is "shaking," EMI-wise, the victim equipment.* The best simulators for this artificial threat are the EFT and ESD tests. Starting with low levels, the injected pulses will be increased until the equipment starts exhibiting malfunctions. If the "FAIL" threshold is significantly *lower than the immunity level that should be expected for this equipment and type of environment,* you will harden one or the other until this installed equipment meets or exceeds the immunity for its category.

Accordingly, over some probation period of a few days or weeks, the user should report a reduction, or disappearance of the problem. The driving assumption being: *If it withstands the standard immunity level on site, this equipment is vaccinated against any kind of short, fast-rising transient, even if the true cause is never found.*

10.9.6   Common Precautions to EFT and ESD On-Site "Forced Crash" Method

The principle of the set-up and routine are basically the same as for development tests, with some noticeable differences in the actual application, considering that you do not have the commodities of a design/development lab. Instead, you are on a customer's site, with probably other equipment still running, people doing their regular work, etc. This implies some precautions, like working off-hours or temporary isolating some section of the plant or building.

(a) There will probably be no ground plane, so you will uses the next, closest metal framework (building girders for instance) as a reference for connecting a temporary aluminum foil. The EFT generator will be grounded to this foil. The EUT will NOT be grounded to the foil, but will keep its normal (presumably safety earth wire) connection to the power panel earth bar.

(b) You are testing an equipment that is actually in use. As the test is progressing, watch for the symptoms that have been noticed by the users. Make sure that your crash evaluation does not lead to material damage or eventually safety hazard. Ask for putting in standby mode the peripheral that could create such a risk.

(c) For EFT, if the EUT power cord does not fit into the generator outlet, use the foil wrapping on the power cord, like for the signal cable (See Sect. 10.6).

10.9.7   Forced Crash Using EFT Simulator

This is, by far, the preferred test to start with, since it requires less time and preparation, being performed on power cables and signal cables only, that are typical weak spots of a system.

Test ALL cables, starting from a low level. For those AE which must functionally operate with the EUT, use ferrite decoupling, as described in Sect. 10.6. The EUT must resist on site, to level that are at least equal to the IEC 61000-4-4 (Fig. 10.8). If these levels are not met, use the fix hints in Sect. 10.6. Some of these fixes may be applicable on the EUT on-site, without too much disruption for the user. Other fixes can make-up for the design weaknesses that cannot be modified on an installed equipment. Those you will apply on the installation itself (Fig. 10.16).

10.9.8   Forced Crash Using ESD simulator

This method is for the cases where there is a suspicion that the problem could be ESD (answer YES in Fig. 10.15 box D). It precisely reveals weak spots on the equipment box and I/O ports area, that an EFT test may have missed. The test procedure is similar to that of lab test, Sect. 10.7. For an installed equipment, the NO-Fail levels of Fig. 10.8 are recommended.

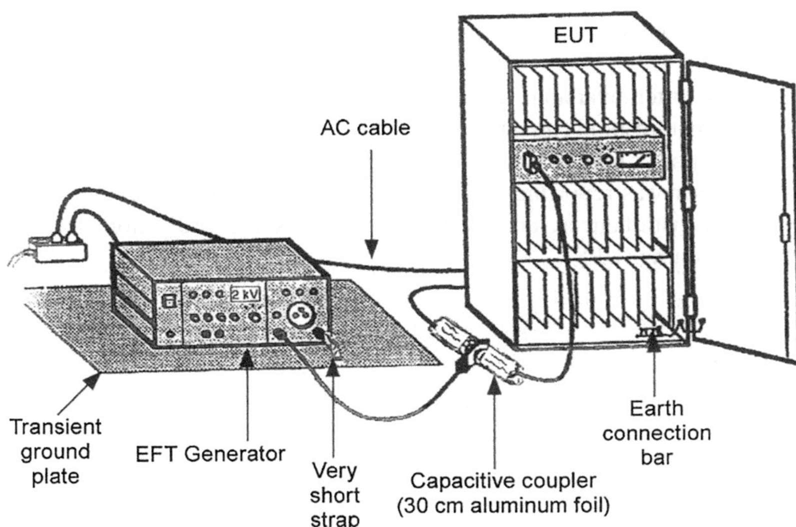

Fig. 10.16 On-site injection of Electrical Fast transients

Warning

This must remain a No-Damage test. Errors are normally tolerated if they are self-recovered, without serious prejudice to the user. Hard errors, especially if they resemble the ones that the EUT user was complaining about are not acceptable. Do not discharge directly on EUT parts that are directly wired to sensitive circuits inside (like connector pins of Digital Bus interfaces, video or RF inputs, etc.), unless the normal factory test of the EUT did include this requirement.

Due to the high voltages (up to 15 kV) developed on the tip of the gun, this method can be dangerous for an equipment whose susceptibility is unknown (not tested by the manufacturer) or even for personnel. Make sure your temporary test area is bound with visible warning tape.

10.10   Power Line Monitoring

There is a domain that an on-site EFT test will not reveal: long-duration power mains disturbances, typically longer than a few tens µs. These long duration pulses are difficult to simulate on-site, because it would require injecting significant disturbances that could affect part or all of the AC distribution at this customer's facility.

So, if occasional power line disturbances are suspected, it is recommended to install a power-line disturbance monitor. This "spy" instrument will survey constantly the short overvoltages and undervoltages, indicating their amplitude, duration, and precise time of occurrence (day, hour, minute). If the sensors are correctly installed, they will show if the disturbance was DM (line-to-line) or CM (Line to Gnd).

Choose, as often available for rental, an equipment with battery back-up, and select a sensing threshold equal to what could exceed the normal (proven) immunity of the EUT to power line. If after some observation period, the EUT had suffered the failures the user was reporting, it will be easy to find if they correlate with some power mains disturbances, and eventually install surge protection devices or a power conditioning (UPS).

Appendix A: EMC Cases Stories

This Appendix is reporting some EMC "war stories" experienced by the author himself or his associates in a US consulting firm. Assisting a customer who has an EMC problem is seldom a boring, "deja-vu" experience. An EMI problem is probably the one whose symptoms can be the most varied and deceptive and whose diagnosis can be the most elusive. Needless to say, EMI is also a privileged playground for Murphy to exercise his laws with demoniac ability.

The following "tales from the trenches" were selected because they share several similarities:

- In general, the plaintiffs were skilled engineers who had tried all common-sense fixes (plus a few others) before calling a consultant.
- The outcome of these fixes (whether or not they helped) usually was not documented or at least not quantitatively. When they were, it was in the form of verbal legacy passed on by each of the frustrated raiders to his next partner. The GO/NO-GO levels reached with each fix, the type, manufacturer, and P/N of the suppression components were not documented. Whether or not the fixes were cumulative or if each fix was taken away before trying a new one could not be determined either. The type of instrumentation used was sometimes not reported. The whole saga summarized as: "We have tried everything, and nothing worked."
- Sometimes, the client had already called a consultant who tried several pieces of the usual EMC arsenal, including in one case an overhaul of the whole building earthing network, ground rods, and so on, which, although it certainly improved the safety of the facility and the wealth of the building contractor, did not do much for the immunity of the system.
- In several instances, many of the unsuccessful fixes were, in fact, appropriate. But they were tried in random sequences so that the engineers could not compile and interpret the gradual changes they might otherwise have noticed, thereby not learning from their mistakes.

M. Mardiguian, *ElectroMagnetic Compatibility*,
https://doi.org/10.1007/978-3-032-02688-0

- Anyway, there was an unanimous consent about the consultant job: He was supposed to fix in 2 days what three successive task forces (appointed by four consecutive division managers) had failed to solve in 8 months.

A.1 Case 1: Spectrometry Equipment

For chemical analysis/detection of metallic pollutants, this system used a 1500 V/40.7 MHz plasma chamber. The HF from the high-voltage cabinet was supplied by two thick 1500 V coaxial cables.

When the company changed from traditional electro-mechanical controls (relays, manual switches, and many more) to a full computer controls, the system went down immediately. The problem was easy to replicate: at the moment the RF power was "ON," the PC collapsed and would not re-start as long as the RF was there.

Analysis The center conductors of the two HV coaxial cables were used as "hot" and "return" conductor, the shields being left floating on the load side!! Asking why such weird scheme was done, we faced one of the die-hard legends EMC consultants are confronted with: some self-proclaimed "expert," at the expense of countless malfunctions that caused havoc in so many systems, had declared that "**cable shields should never be grounded at both ends**, because this create ground loops".

The funny side was that this odd configuration of the HV coax cables had lasted for years w/o problem, as long as the control was a rugged electro-mechanical gear. Things went berzerk when a computer control was installed.

An interesting question arises: which way(s) was the HF return current using? Given that current always looks for all possible impedance paths, it seems possible that some % of current might have seeked for alternatives to the hair-pin trip of the two coaxial cables center wire.

Such alternative paths were offered by the chamber chassis and the general grounding network of the entire system (Fig. A.1). And indeed, using a current

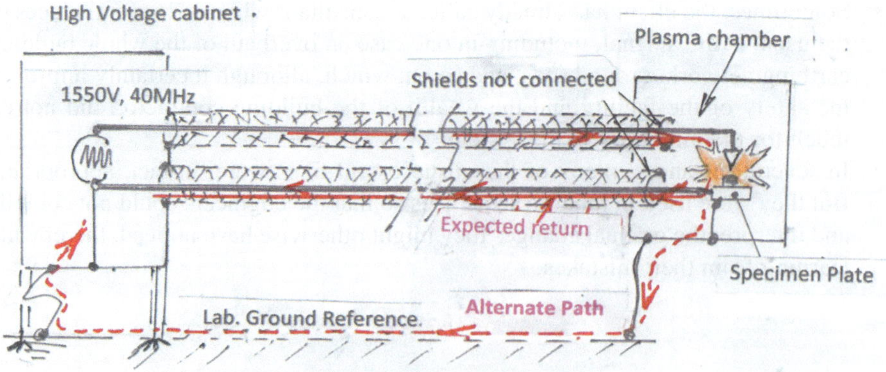

Fig. A.1 Configuration of the Hi-voltage/40 MHz initial cabling to the plasma chamber

probe, 40.7 MHz currents were found on various metal members of the system, creating local strong magnetic fields and ground shifts along the elements of the new electronic controls.

We simply re-wired the HV coax such as the HF current would return to its source via the shield, which is the normal, low impedance path, making the problem disappear forever.

A.2   Case 2: The Trigger-Happy Airbag Initiator

This case which happened in a car manufacturing plant is a not so rare situation where an upset manifests only when another, un-related, extraneous event exist.

As the production of a new model had just started, few air-bag incidents were reported on the first produced cars. In one case, the vehicle was parked, engine running at idle, the driver simply wiping the dashboard with a cloth, when the airbag triggered unexpectedly. The sudden blast of an airbag, especially when the driver or passenger are not normally seated and belted, is a painful experience: The life-saving device turns into deafening, brutal punch. It had been a dry, cold winter day, so an ESD was immediately suspected, and a solution urgently needed since the delivery of the vehicles to the car shops was ready to start.

For helping a fast and efficient investigation, a complete car harnessing with all the sensors, connectors, modules, and so on, was laid on a work bench, powered-up by a battery. Airbag EED initiators were in place, with dummy charges. All arc-reachable points were patiently tested, including those where the gun had to be cranked up to 30 kV to get an arc creeping to hidden connector pins. No firing ever occurred.

With the entire product engineering on emergency mode, a full support of all concerned parties was available, several investigations being run in parallel. After few days, interesting clues emerged from reports coming in the war-room:

- first, it was found that an improper assembly-line routine resulted in one wire of the airbag harness being sometimes pinched under the front seat steel armature. The random grounding of this wire did not result in a solid short, and could go un-noticed, but it caused a differential pair, balanced link to become a single-ended one, with a strong degradation of its Common Mode Rejection.
- then, a precise re-creation of the event revealed that, at the moment of the incident, the car heating was on full "Warm." The airbag control module happened to be near the foot of the steering column, close to the outlet of the forced hot air. The temperature inside the module was not pushing the components beyond their limits, but the noise margin of the digital gates and accelerometers comparators was seriously reduced.

When those two adverse conditions were artificially recreated, the EED instantly triggered for ESD levels as low as 5–6 kV. One immediate corrective action was to review the assembly procedure and add an insulation test on this branch of the airbag harness; later, a relocation of the airbag control module was scheduled.

A.3 Case 3: RF Emissions from Ship's Radar Disabling Hundreds of Cars Parked on the Pier Feb 2000, Santa Barbara (CA, USA)

As the aircraft carrier USS Lincoln was approaching the harbor, heading for its shipyard, many cars parked in zones as far as 1.8 mi were stuck on place, the owners being unable to unlock the doors or start their engines. Investigations led to believe that the powerful radar and Radio transmitters were jamming the cars electronic devices, which were not protected.

March 2001, Bremerton (Wash, USA). A similar incident occurred as the USS Carl Vinson entered the harbor. Garages were swamped by hundreds calls from furious owners who could not lock, unlock or start their vehicle.

Analysis Out of any official mission, we reviewed the problem for checking the likelihood of a true EMC issue.

(a) **The source** was known, but could not be blamed or modified: a battleship, wherever it may be, keeps its radar and essential radio-com in full readiness (probably lessons learnt from P. Harbor).

(b) **The victim's susceptibility to RF:** since early 90s, car electronics are designed and tested to be insensitive to the most severe ambients like proximity of powerful RF transmitters. Car manufacturers started testing safety-critical, essential devices for an immunity to 100 V/m, some of them even 200 V/m.

(c) **The coupling path**: could, and how far, the ship radar and most powerful RF transmitters exceed such values?

Frequencies and RF power of such Navy transmitters are confidential, so we used typical high "default values": 1 MW transmitter with 34–36 dB gain for directional beam.

In free, lossless propagation, the calculated received field is (Fig. A.2):

>200 V/m up to 250 m distance
>100 V/m up to 500 m distance

Therefore automobile devices could possibly be disturbed as far 250–500 m, depending on the actual level of their EMC evaluation.

Yet the question remains: how come that some cars were disturbed as far as 1.8 mi away, which is six times further than their expected frontier of malfunction?

Asked about this, the responses of car manufacturer representatives were:

(a) "It cannot be EMI: lock/unlock control uses a personal owner code: it is impossible that an RF ambient could replicate the code." *The truth*: Immunity tests were carried as follows: "*Under strong RF field, car should not lock—unlock by itself.*" But nobody checked if the owner could open or lock his car while it was illuminated by RF! The actual failure mode was that the receiver circuit was saturated by the strong, rectified RF field, and could not recognize ANY signal. But this scenario was not tested.

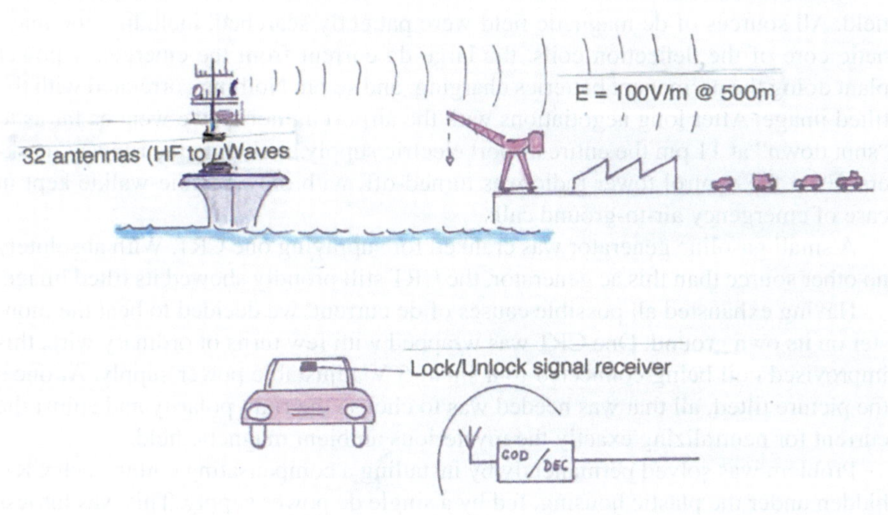

Fig. A.2 Radiated emissions from battleship jamming car lock/unlock device

(b) Other manufacturer: "This car model uses an infra-red transmission. It is immune to EMI" *Wrong*: the module is not optically disturbed. The RF upset takes place on the electric side of the photoelectric conversion. The RF carrier is captured by some internal wiring, including power supply, is demodulated, and saturates the first stage of the RF amplifier.

Lesson Learnt Do not trust blindly EMC tests results until the real test procedure and accept/reject criteria have been checked. The cars that suffered this episode had not been correctly tested for immunity to intense RF fields in a real situation.

A.4   Case 4: Disturbed Air-Traffic Control, Nice Airport (1991)

By its traffic, the Nice airport ranks third in France. Its newly erected control tower was 38 m high, with last floor occupied by traffic control agents supervising take-off, approach and landing of national and international flights.

From the very beginning, the eight CRTs attended by the ATC agents had suffered from distorted image, making the watching very tiresome. The airport technical staff, suspecting a magnetic cause called for our assistance.

Picture distorsion of CRT was common place before the flat LCD displays era, so we came in with the biased idea that this was a case of close proximity with building power wiring. Immediately, a look at the distorted image showed that it was something else: picture was neat, but heavily slanted (30–45°) on one side. Moreover, of the eight tubes arranged in circle, one half had the tilt on the left, the other four tubes showing a tilt the other way.

The cause was obviously not an AC mains current effect, but a stable dc magnetic field. All sources of dc magnetic field were patiently searched, including the magnetic core of the deflection coils, the large dc current from the emergency power plant doing the permanent batteries charging, and so on. Nothing correlated with the tilted image. After long negotiations with the airport authority, we went as far as to "shut down" at 11 pm the entire airport electric supply, including the runway markers. Even the control tower radio was turned-off, with only a talkie-walkie kept in case of emergency air-to-ground calls.

A small gasoline generator was cranked for supplying one CRT. With absolutely no other source than this ac generator, the CRT still proudly showed its tilted image.

Having exhausted all possible causes of dc current, we decided to beat the monster on its own ground. One CRT was wrapped with few turns of ordinary wire, this improvised coil being connected to a small 5 V adjustable power supply. At once, the picture tilted, all that was needed was to choose the right polarity and adjust the current for neutralizing exactly the mysterious ambient magnetic field.

Problem was solved permanently by installing a compensating coil in each CRT, hidden under the plastic housing, fed by a single dc power supply. This was far less expensive than an integral thick magnetic shield, made of high permeability metal, cut and formed to size, which had been also foreseen.

But a frustration remained: what was the cause? Only speculations were possible, of which we kept the following two:

- Being a high structure with metallic skeleton, the tower is a target for frequent lightning strikes. French Riviera has one of the highest lightning records in France, so the tower was correctly protected by lightning rods. Yet 10–100 kA currents running from the top to ground could have permanently magnetized the tower. This remanent H-field was a likely cause.
- During the tower construction, steel beams were hoisted by a crane using a magnetic plate for holding the loads. These beams had kept a remanent magnetic field.

In either case, the drastic solution would have been to de-magnetize the tower …. A gigantic task and probably useless if the lighting assumption was the right one.

Lesson Learnt   "Whatever works …".

A.5   Case 5: Switched-Mode Converter Module

Violation >15 dB of radiated emission limit because of leaky cover lid.

Our assistance was required for an airborne equipment whose pre-production model was exceeding the most severe RE-102 limit, re-inforced by 10 dB. The engineering team was EMC competent and had already tried classical fixes, with no or feeble success.

The EST was a dc–ac inverter, fully enclosed in a cast aluminum box without openings, except the I/O port, equipped with shielded connectors and shielded

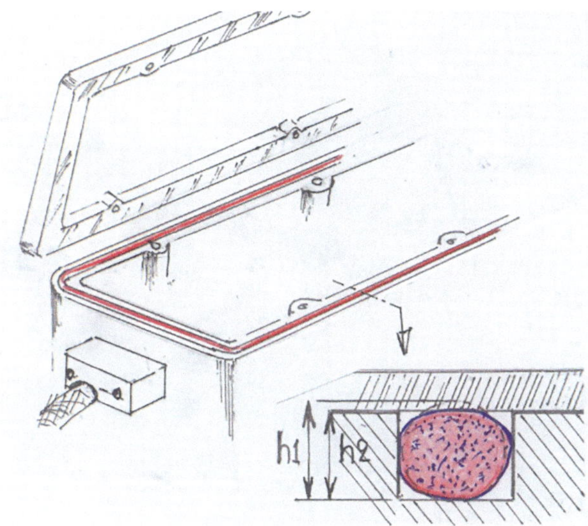

Fig. A.3 Gasket mount in its groove: the difference $\Delta(h1 - h2)$ was too small for good gasket compression

cables. Limit violation was significant (Fig. A.3) in the 30–200 MHz, which is about the domain where the radiating element could be the EST box itself or the external cable. This latter was rapidly segregated by a heavy ferrite loading (three toroids with two turns each) close to the I/O ports: the Off-spec levels were almost unaffected.

Examination of the cast aluminum cover lid showed finely ground mating surfaces, with an EMC gasket inserted in a machined groove. The gasket selected by the engineers seemed adequate, with a claimed 80 dB shielding effectiveness. However, having sometimes got our fingers burnt by gaskets vendors' data, we insisted for performing the *"cigarette paper test"*:

After wiping the mating surfaces clean of grease or finger traces, a very thin paper strip is inserted in few selected places, half-way between two cover screws. Once the cover screws are tightened at the proper couple, it should not be possible to pull-out stripes without tearing them-off. This would be a proof that the gasket is well compressed for a good lid-to-box contact. On this EUT, the paper strips were easily taken away, without any effort. Obviously the combination gasket-to-groove dimensions were not adequate (Fig. A.3).

Unless having a full inventory of gaskets with various sizes and hardness, it seemed difficult to validate quickly the deficiency of the gasket-to-lid contact. So a very simple fix was tried: the gasket was removed, temporary replaced by a three-fold sheet of aluminum foil (kitchen type, 15 µm thick). There was no need to cut-out the inner part, since what counts is a decent contact between lid and box edges. With a triple thickness of 45 µm, the pressure of the mounting screws was enough to grant a good contact.

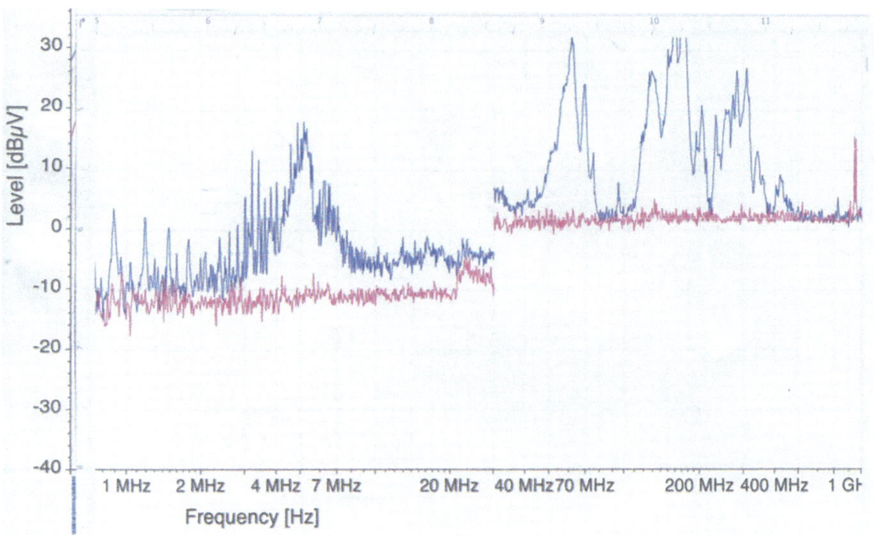

Fig. A.4 Radiated E-Field of the box alone. The **Blue trace** is the box with initial EMC gasket. **Red trace**: after replacing the EMC gasket by 3 layers of 15 μm aluminum foil, the RE level barely exceeds the background noise. The sudden 30 MHz step is due to the change of receiver bandwidth. Vertical scale is the output signal (dBμV) from the measuring antenna, since what was looked for were the relative improvements

Figure A.4 shows that with this aluminum foil temporary gasket, the initial radiated spectrum went down to almost the measurement background noise (red trace), an evidence that the present gasket was totally useless.

We recommended that a decent gasket be selected, with a careful checking that the groove tolerances were correctly matching with the gasket diameter for a $\approx 15\%$ to 25% compression, including mechanical tolerances when mating faces were just touching.

Having solved the box leakage problem, what remained to improve was the contribution of the shielded cable exit. This was relatively easy once the box leakage has been fixed; a more careful bonding of the shielded I/O receptacle to the box face was validated, and the whole EUT was compliant.

A.6   Conclusion: Troubleshooting Hints

Armed with the explanation of coupling mechanisms given throughout this book, reader should be able to identify most of the EMC failure modes and apply the proper fixes. To help quantifying the improvements, and avoiding false routes, the following guidelines have proven to be useful:

- *Always grade the progress* brought by a modification in terms of the new Run/ Fail levels, or, for pulse tests, in terms of the new error-per-pulse ratio (see Chap. 10 Sect. 10.6.3 and [13]).
- *Try to visualize the parasitic current paths* and how they are modified by bonding/grounding changes. This often gives a clue to the validity of a fix.
- *Monitor either the ground currents or the common-mode current* induced in cables. For this an RF current probe is a priceless tool. Being a totally floated sensor, it is not affected itself by EMI, and it does not modify the victim circuit.
- As an *absolute rule, never remove a fix* which, although it seemed to make sense, did not bring any significant improvement. Noise can couple into the victim by many parallel paths, and until ALL of them have been reduced, a change may not be seen.

Last *lesson learnt*: even if a root cause is obvious, an EMI problem may sometimes manifests only when it is accompanied by another environmental event. When it seems impossible to re-create the problem by the "forced crash" method, look-out for a possible joint effect of temperature, vibration, power supply undervoltage or overvoltage, earth or ground voltage gradient, etc.

Appendix B: A Few Simple EMC Software Programs

Here we describe simple, comprehensive software models, which calculate quickly the effect of four frequent EMI mechanisms:

- the attenuation of a filter for high speed logic signals,
- the coupling of a common mode voltage applied to an unshielded twisted pair (UTP),
- the coupling of a common mode voltage applied to a coaxial cable,
- the Shielding effectiveness of a box with a given aperture.

These software are built on SPICE programming (Simulation Program with IC Emphasis), which can be used for much more than just Integrated Circuit design, and has been favored by many EMC engineers as a very practical, handy tool for EMC analysis and prediction.

The version of SPICE used here is an early one, that we personally prefer because it is simpler than more recent releases, allowing the user to freely choose its own numbering of nodes, components etc., which we find more friendly than more sophisticated versions.

B.1   Attenuation of a Filter for High Speed Digital Signals

Compared to power line filters, signal line filtering is a different challenge: signal filters must keep the desired signal unaffected in amplitude and phase, from dc up to the highest bandwidth required for signal integrity. If not given by the datasheet of the signal technology, this useful bandwidth can be derived from the clock frequency, or the rise time of the fastest signal.

© The Editor(s) (if applicable) and The Author(s), under exclusive license to
Springer Nature Switzerland AG 2025
M. Mardiguian, *ElectroMagnetic Compatibility*,
https://doi.org/10.1007/978-3-032-02688-0

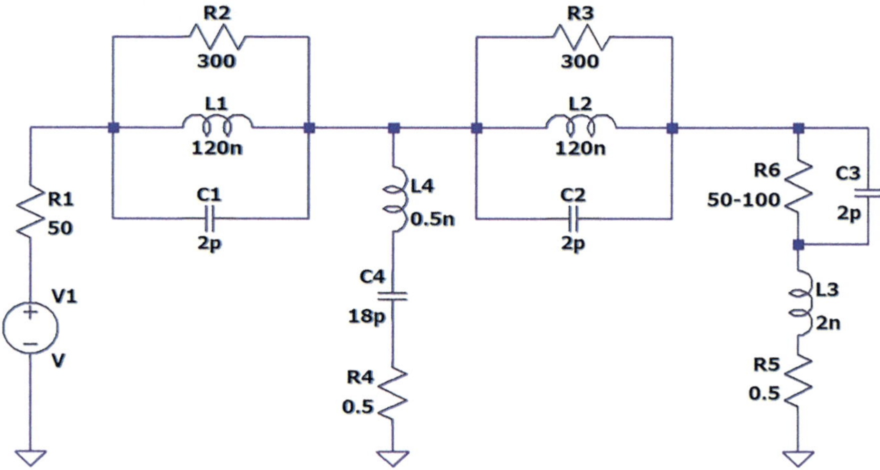

Fig. B.1 Schematic for the signal filter of the example, including parasitic elements

The cut-off frequency of the filter is derived from the useful bandwidth.
The user must define the desired attenuation, and at which frequency or
frequency band.

With these figures at hand, the number of elements of the filter is determined.

The impedances on the source and load side of the filter must be known to opti-
mize the arrangement of the filter L, C élements. Details on filter selection are given
in Chap. 8, Sect. 8.5.

All these data are entered in the SPICE program for calculating the filter attenuation (more exactly the insertion loss). The filter schematic is described in Fig. B.1,
including its parasitic elements. These elements are causing the actual attenuation
(curve B on Fig. B.2) to depart significantly from that of a filter with ideal components (curve A). Values used for parasitics are typical default values for discrete,
PCB-mounted components.

B.2 Coupling of a Common Mode Voltage onto an Unshielded Link

This program calculates the differential voltage on the load end of a single unshielded
tw. wire pair (UTP), when submitted to a common mode (CM) voltage. This CM
voltage may be the result of either:

- A conducted phenomena: Common Impedance Coupling, as described in
  Chap. 2.
- A radiated, field-to-cable coupling as described in Chap. 3.

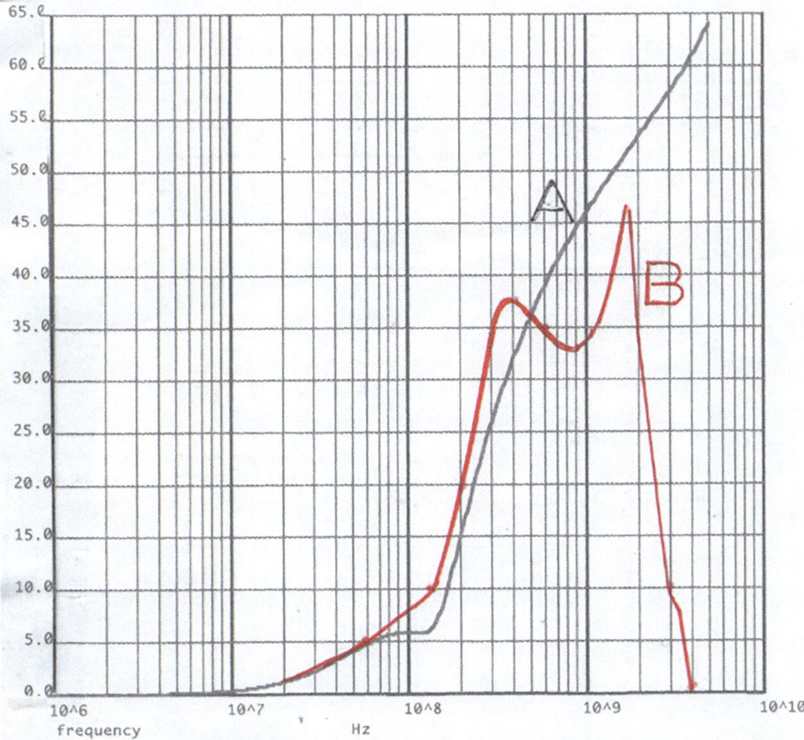

Fig. B.2 Filter for high speed signals (useful bandwidth 35 MHz). With the "ideal" filter, attenuation (**A**) is skyrocketing, while actual filter curve (**B**), taking into account parasitics, reveals a more realistic figure

In both cases, the resulting voltage is injecting CM currents in the two wires of the pair. What is of interest is the resulting differential voltage sensed by the victim's input impedance. The program gives the Conversion Coefficient $Kr = V_{(victim)}/V_{(CM)}$.

The example below gives (Fig. B.3) this Mode Conversion as: **$Kr(dB) = 20\ Log$ $V_{(victim)}/V_{(CM)}$** for a 3 m link, with two configurations for the victim's circuit: *isolated* from the local (chassis or earth) ground or with signal *0 V reference grounded*. While it is dc-isolated, the victim's reference is still virtually grounded at HF by its PCB stray capacitance Cp.

It is clear that floating the PCB 0 V vs. the local chassis or earth ground gives a significant advantage at low frequencies, below the megahertz. But this advantage vanishes above a few megahertz, making this presumed isolation worthless or even *detrimental around 8–30 MHz*, where the differential voltage seen by the victim is even worse than with the grounded version.

Details of the SPICE simulation. The 3 m transmission line is simulated by five 0.60 m segments, each with 60 mΩ/0.6 μH/35 pF pair capacitance, and 6 pF wire-to-gnd capacitance (Fig. B.4).

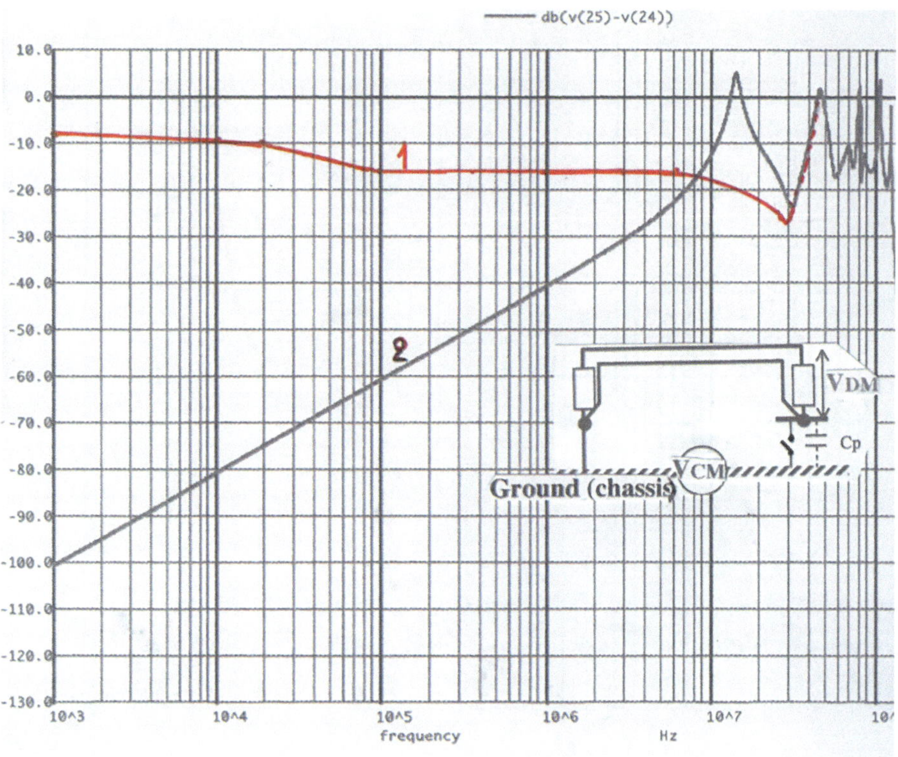

Fig. B.3 Calculated CM-to-DM conversion Kr for a 3 m link with UTP, where the victim's 0 V reference is: (1) connected to the local ground or (2) floated from the local ground

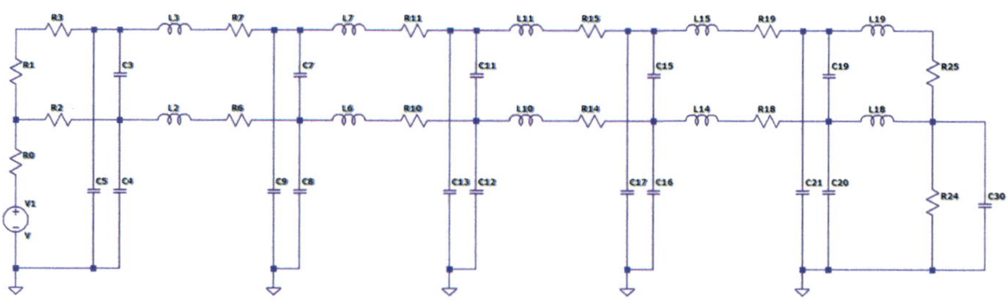

Fig. B.4 Schematic used for the SPICE program Unsh. Twisted pair

B.2.1   Breakdown of the Program Instructions

In the last segment, isolation transformer or opto-isolator can be inserted before the load resistance R25, for improving the CM immunity.

```
V1 1 0 0 AC 1
R0 1 2 1m
*(if driver side floating)
*C00 1 2 200p
R01 2 3 100
* Txm line 5 segments 0.60m
R2 2 4 60m
R3 3 5 60m
L2 4 6 0.6u
L3 5 7 0.6u
K2 L2 L3 0.7
C3 5 4 35p
C4 4 0 6p
C5 5 0 6p

R6 6 8 60m
R7 7 9 60m
L6 8 10 0.6u
L7 9 11 0.6u
K6 L6 L7 0.7
C7 9 8 35p
C8 8 0 6p
C9 9 0 6p

R10 10 12 60m
R11 11 13 60m
L10 12 14 0.6u
L11 13 15 0.6u
K10 L10 L11 0.7
C11 12 13 35p
C12 12 0 6p
C13 13 0 6p

R14 14 16 60m
R15 15 17 60m
```

```
L14 16 18 0.6u
L15 17 19 0.6u
K14 L14 L15 0.7
C15 16 17 35p
C16 16 0 5p
C17 17 0 5p

R18 18 20 60m
R19 19 21 60m
L18 20 22 0.6u
L19 21 23 0.6u
K18 L18 L19 0.7
C19 20 21 35p
C20 20 0 5p
C21 21 0 5p

* Normal, un-filtered Termination (Victim side)
R27 23 25 1m
R25 25 24 100
C25 25 24 1p
R26 22 24 1m

*C30,R30: PCB-to-gnd Parasitic cap // to 0v-chassis
resist. if PCB grounded
C30 24 0 30p
R30 24 0 10Meg

.AC DEC 400 1k 300Meg
.END
.control
run
plot db(v(25)-v(24))
.endc
```

B.3   Coupling of a Common Mode Voltage onto a Coaxial Cable

When one needs to know the efficiency of a shielded cable, the user is often seeking for a shielding effectiveness (SE) figure, because it can relate the cable to the effectiveness of housings (boxes, cabinets, etc.) of his system.

Unfortunately, SE of a shielded cable is a versatile figure, since it depends not only on pure cable parameters (diameter, type, and structure of shield material) but also on external variables which are installation-dependent: height of cable above ground, terminating impedances at each end, nature of the incident EM field: far field or near-field, and in this latter case, is it electric dominant or magnetic dominant? etc. Therefore, as explained in Chap. 5, Sect. 5.3, the real SE of a cable is hard to get, depending largely on factors which are not part of the cable physical elements.

The following program is using some well-established jacket shield elements, generally present in the transfer impedance Zt, and combine them with cable installation factors like height above ground, shield connections to ground and source/ load impedances. Like for the previous single pair case, the SPICE result is the CM-to-DM coupling coefficient $\mathbf{Kr(dB) = 20\ Log\ V_{(victim)}/V_{(CM)}}$.

The calculation example (Fig. B.5) is made for a 5 m ordinary single braid coaxial cable (RG58). It can be easily tailored to any shielded cable whose physical elements are known. When the shield is grounded via a tight, circular contact (BNC for inst.) the reduction factor Kr improves rapidly with frequency, stabilizing at 50–55 dB from 0.5 to 15 MHz. Beyond this frequency, for the 5 m length, mismatch

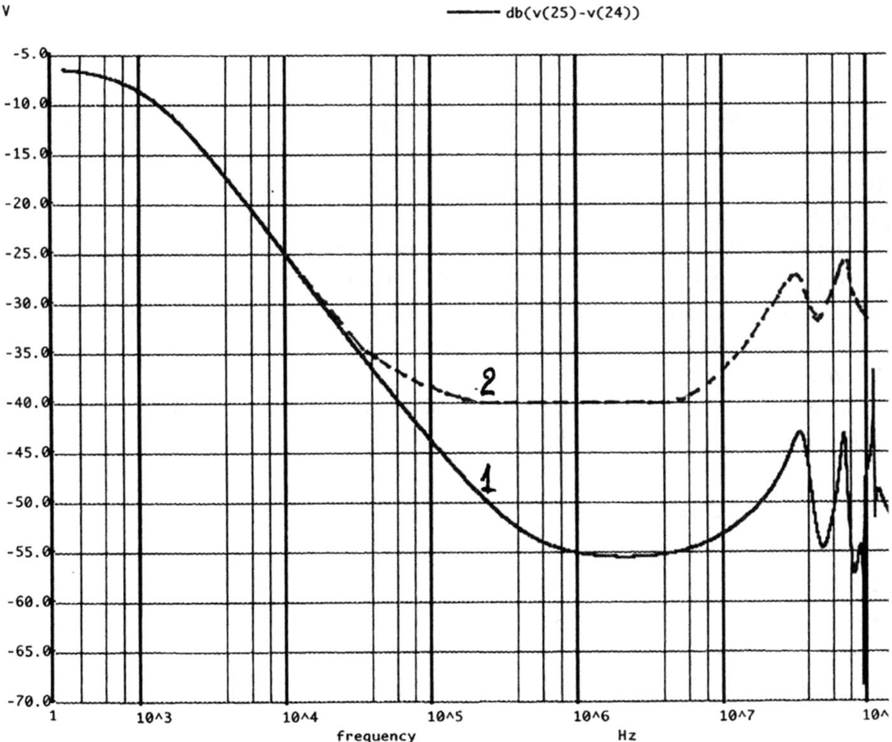

Fig. B.5 Single shield coaxial cable, RG58 type. Calculated CM-to-DM conversion Kr for a 3 m link, (**1**) normal grounding of braid by 360° connector backshell. (**2**) Victim's end of shield grounded by a 10 cm pigtail

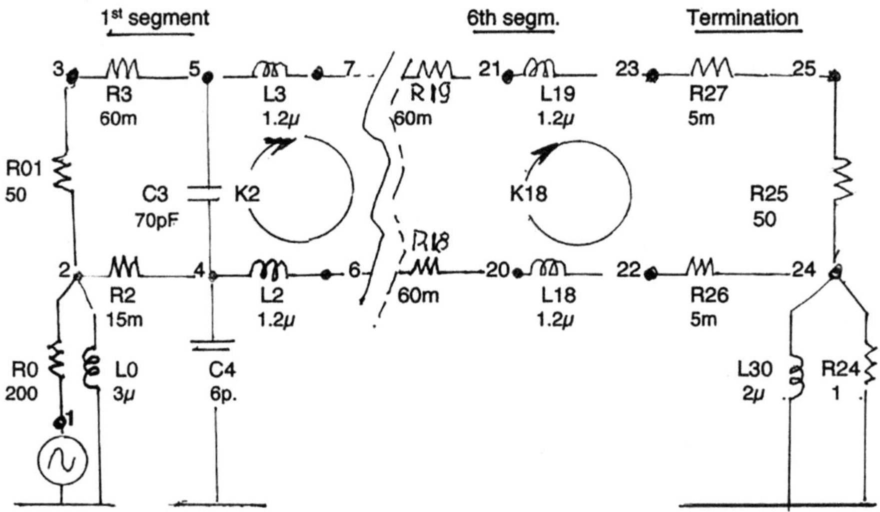

Fig. B.6 *Partial* view of schematic used for the SPICE program 5 m coaxial cable. Only the first and last segm

ringing start to show-up, causing periodic 5–10 dB drops of Kr. But even with this effect, the protection of a single shield coaxial remains excellent: 45 dB ≈200 times reduction of the CM coupling for the victim end. If the shield is grounded via a 10 cm pigtail, a disastrous—but still frequent mistake—the transfer impedance is severely spoiled, causing a 15 dB deterioration of Kr (Fig. B.6).

Details of the SPICE program for coaxial cable Comm.Mode ⟶ Diff. Mode conversion (K-factor per Chap. 5 Sect 5.3). Instructions are somewhat similar to these of a single pair, "Coupling of a Common Mode Voltage onto an Unshielded Link" section. The Leakage transfer inductance is simulated by mutual coupling **$K = 0.995$.**

Shield data: Rshield: 15 mΩ/m, Lt(leakage) inductance: translated into mutual inductance (shield-to-core) KL2-L3, etc.

```
V1 1 0 0 AC 1
R0 1 2 200
L0 1 2 3u
*(si flotté côté génér)
*C00 1 2 200p
R01 2 3 50

* Txmission line segments ( 5 x 1m)
R2 2 4 15m
R3 3 5 60m
L2 4 6 1.2u
```

```
L3  5  7  1.2u
K2  L2  L3  0.995
C3  5  4  70p
C4  4  0  6p
*C5  5  0  6p


_ _ _ _ _
  _ _ _ _ _ _ _
* 6th (last) segment
R18  18  20  15m
R19  19  21  60m
L18  20  22  1.2u
L19  21  23  1.2u
K18  L18  L19  0.995
C19  20  21  70p
C20  20  0  6p
*C21  21  0  6p


* Normal shield Termination via  connector Zt
R27  23  25  5m
R25  25  24  50
*C25  25  24  10p
R26  22  24  5m
L30  24  0  2u
R30  24  0  1


.AC DEC 400 300 150Meg
.END
.control
run
plot db(v(25)-v(24))
.endc
```


B.4 Shielding Effiectiveness (SE) of an Enclosure with a Known Aperture

Based on a simplified, but validated model, the program calculate de SE of a metallic (or simply conductive) box with a given aperture. The basic theory is explained in Chap. 6, Sects. 6.3 and 6.4.

The incident field is simulated by a voltage source in series with a 377 Ω internal resistance, forcing a far field condition, which is a radiating source further than $\lambda/2\pi$ from the box.

- *The aperture* is assimilated to an inductance, whose impedance at lower frequencies is bound by the dc resistance of the surrounding metal, increasing with frequency, until reaching the point where it resonates with the slot edge-to-edge capacity. This can be seen as the reverse of a dipole.
  Aperture inductance varies linearly with its length (or diameter if it is a round hole) and with the logarithm of its height if it is a rectangular slot.
- *The volume behind the slot* (the inside of the box) being not an infinite space, the program accounts for the reflection of the field on the opposite wall, bouncing back and forth toward the input aperture, with a calculated phase shift. This represents the self-resonance of the box alone: at each box resonant frequency, there is an increase of the internal field, causing a sharp drop of SE value.
  To mimic the progression of the field after penetrating the box, up to the opposite face, the box length is seen as a transmission line with distributed inductance, capacitance and resistive losses. This is simulated by a cascade of six 10 cm segments, which is an acceptable trade-off up to the first resonance of the slot.

The exact resonance of the aperture is competing with those of the box, with a SE peaking down to 0 dB. The program stops there, since beyond that point, the multiple SE hops from zero to high values offer no interest.

The "Q" of the cavity, which is high for an empty box, is damped by affecting a resistive load at the end, which correspond to a more actual behavior of an electronic box full of PCBs, cables, components, etc.

B.4.1   Two Interesting Options Are Offered by the Program

- In the set of input values, next to the 377 Ω line, a 1.25 µH inductance can be used in cases where the source is a predominent H field type. This will force a field impedance <377 Ω up to a frequency where the inductor impedance reaches 377 Ω, retrieving the far field conditions. This 1.25 µH can be activated or deactivated by simply adding an asterix on this SPICE input line.
- The in-box location where the exact value of internal field is needed can be selected by asking for the node voltage at any point along the line.

Figure B.7 shows an example of the program result applied to a 60 cm long box, with a single 10 × 1 cm aperture. Plot A is the SE for a target near the box far end. Plot B is the SE mid-way. The first downfall of SE at 250 MHz is due to the first resonance of the box alone, excited by the radiation from the slot. Then the cascade of peaks and valley reveals the periodic reflections of the box as a mismatched "transmission line." Finally the true resonance of the slot is seen as a 0 dB SE around 1000 MHz.

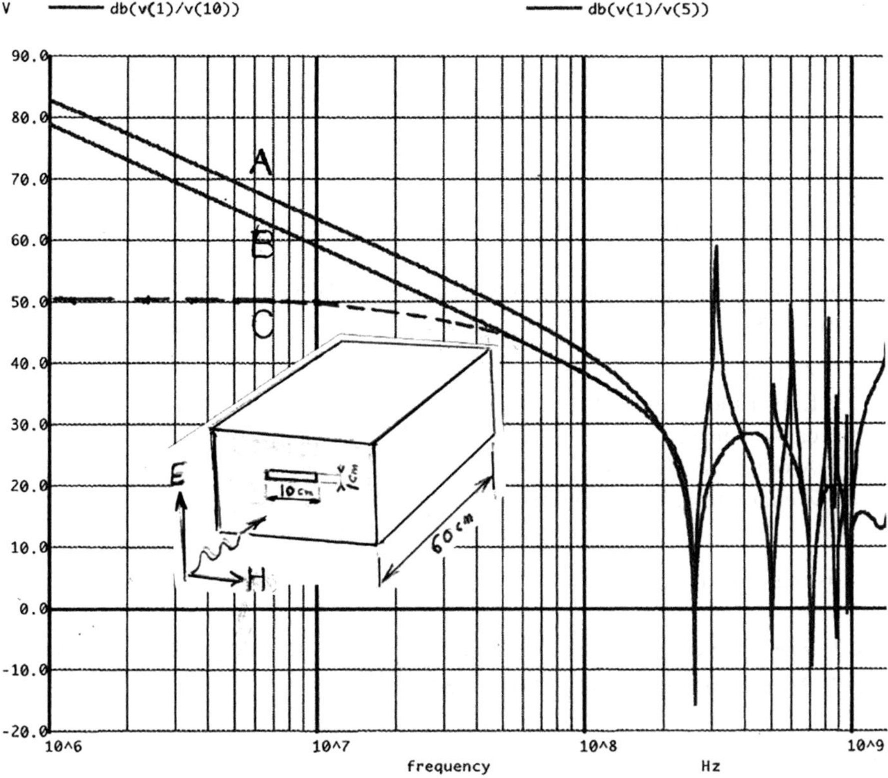

Fig. B.7 Result of the SE program for a 60 cm long box, with a **10 × 1 cm aperture**. (**A**) Field at the far end of the box, (**B**) field at a point midway. (**C**) Case of a H-field source, 1 m from the box façade

Plot C is the calculated SE if the radiation was from a H-field source at 1 m distance: it starts with a flat portion, since in near field condition the wave impedance is $\ll 377\ \Omega$, increasing progressively with frequency, while the slot impedance is also increasing with frequency, resulting in a constant value of the aperture reflection loss. Around 48 MHz, which is the frontier of the near-far field transition for a 1 m distance, the SE is smoothly rejoining the far-field SE curves.

B.4.2 Useful Data

• Self-inductance and capacitance of a slot:

	Length-to-height ratio				
	10	30	100	300	1000
L(slot), nH/cm:	1	0.7	0.5	0.4	0.3
C (slot), pF/cm:	0.1	0.15	0.2	0.3	0.4

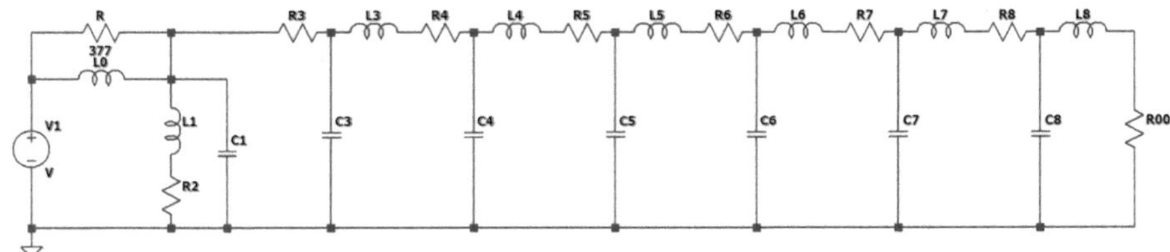

Fig. B.8 Schematic used for the SPICE program of the box with 10 × 1 cm slot

- Distributed elements for the box treated as a Wave guide excited by the slot (Fig. B.8)

$$Z \text{ carat} = \sqrt{(1 - \lambda/2a)^2}$$

Linear inductance: 0; 1 μH/m.

capacitance : 1 pF/m

Around cut-off frequency: $Zc \approx 328\ \Omega$

Z charact. = 377 (h/w) with w: box width, h: box height

Linear inductance: 0.1 μH/m

Listing of instructions for the SE box program:

```
*LO simulates conditions of a near H field, for d=1m
*if d≠1m, make LO=1.25u x d

V1 1 0 0 AC 1
R1 1 2 377
*LO 1 2 1.25u

* slot: 1nH/m, 1pF/m , for  1/h =10
C1 2 0 1p
L1 2 3 10n
R2 3 0 5m

* Transm line equivalt to box. 10cm segmnts
R3 2 4 100m
L3 4 5 100n
C3 4 0 1p

R4 5 6 100m
L4 6 7 100n
C4 6 0 1p

R5 7 8 100m
L5 8 9 100n
C5 8 0 1p
```

```
R6 9 10 100m
L6 10 11 100n
C6 10 0 1p

R7 11 12 100m
L7 12 13 100n
C7 12 0 1p

R8 13 14 100m
L8 14 15 100n
C8 14 0 1p

R00 15 0 1
.AC DEC 200 1meg 2000meg
.PRINT AC Vdb(1) VP(1)
.END
.control
run
plot db(v(1)/v(5))  db(v(1)/v(10))
.endc
```

Correct (or Best) Answers to Quiz
Chapter 1

1.   (b)
2.   (c)
3.   (b)
4.   (a)
5.   (c)

Chapter 2

1.   (d)
2.   (a)
3.   (d)
4.   (d)
5.   (d)

Chapter 3

1.   (b)
2.   (c)
3.   (c)
4.   (c)
5.   (b)

Chapter 4

1.	(d)
2.	(a)
3.	(d)
4.	(b)
5.	(b)

Chapter 5

1.	(a)
2.	(d)
3.	(b)
4.	(d, 20–40 mV)

Chapter 6

1.	(d)
2.	(c)
3.	(c)
4.	(b)
5.	(b)
6.	(b)

Chapter 7

1.	(c)
2.	(c)
3.	(c)
4.	(b)
5.	(a)

Chapter 8

1.	(d)
2.	(a)
3.	(b)
4.	(c)
5.	(c)

Chapter 9

1.	(b)
2.	(a)
3.	(b)

Glossary of Essential Terms and Symbols

λ Wavelength in air

$\lambda(\mathbf{m})$ 300/F(MHz)

ω Angular frequency = 2π F

BB, NB BroadBand, NarrowBand: when only one (NB) or many harmonics (BB) are seen by the receiver

BCI Bulk Current Injection

CE, CS, RE, RS Conducted Emission, Susceptibility, Radiated Emission, Susceptibility

CW Continuous Wave (most generally sine wave, like a RF carrier)

DM, CM currents or voltages Differential Mode (wire-to-wire), Common Mode (wires vs. ground)

EUT Equipment Under Test

(N)EMP (Nuclear) Electro Magnetic Pulse

VHF, UHF, SHF Very High (30–300 MHz), Ultra-High (300–3000 MHz) or Super High (3–30 GHz) Frequ.

M. Mardiguian, *ElectroMagnetic Compatibility*,
https://doi.org/10.1007/978-3-032-02688-0

References

1. Hoeft, L. Transfer Impedance of backshells, 1982 IEEE/EMC Symposium
2. Mardiguian, M. Prediction method for cable shielding factor, based on Zt, ITEM, 2012 Design guide
3. Mardiguian, M. Differential Txfer Impedance of shielded twisted pairs -ITEM, 2010 Design guide
4. Palmgreen, C. Shielded Flat Cables for EMI/ESD reduction, 1981 IEEE/EMC Symposium
5. Schelkunoff, S. Electromagn. Theory of coaxial lines and cylindrical shells, Bell Technical Journal, 1934
6. Vance, E. Coupling to Shielded cables, Wiley, 1978
7. Leferink, F. Shielding Basics, IEE/EMC Symposium, 2010, Ft Lauderdale
8. Casey, F. Shielding of wire-mesh screens, IEE/EMC Transactions, Aug 1980 Vol 30
9. S. Caniggia, S. "Shielding of Enclosure" Commitato Electronico Italiano, Aug 2018
10. Ott, H. Electromagnetic Compatibility Engineering, Wiley, 2009
11. White, D.R.J, Mardiguian, M. Electromagnetic shielding, ICT,Gainesville, VA, 1988
12. Mardiguian, M. Controlling Radiated Emissions", Springer NY, 2014
13. M. Mardiguian, "EMI Troubleshooting Techniques" (McGraw Hill, 2000)
14. H. Ott. "Workbench EMC measurements" (www.hottconsultants.com)
15. Wyatt, K "Troubleshoot Radiated Emissions" Interference Technology ITEM (2010)
16. Mardiguian, M. ESD Understand, Simulate and Fix, Wiley, 3rd edition 2009
17. IEC 61000-4-2 ElectroStatic Discharge Immunity Test (2008)
18. ANSI C63-16 Standard for ESD Test Methodology and criteria (2005)
19. Calcavecchio, R. A standard test to determine ESD susceptibility. IEEE/EMC sympos. 1986
20. Boxleitner, W. ESD and Electronic equipment, IEEE Press, 1989
21. King, M. "Mastering ESD System response". EMC Technology Magazine, March & May 1988
22. Mardiguian M, Caron-Fellens J.P, The Intelligent Concrete ®, Economical Technique for Architectural Shielding. EMC Society Magazine Jan 2018

Index

© The Editor(s) (if applicable) and The Author(s), under exclusive license to
Springer Nature Switzerland AG 2025
M. Mardiguian, *ElectroMagnetic Compatibility*,
https://doi.org/10.1007/978-3-032-02688-0